Veröffentlichungen des Instituts
für Deutsches, Europäisches und Internationales Medizinrecht,
Gesundheitsrecht und Bioethik
der Universitäten Heidelberg und Mannheim   33

Herausgegeben von
Peter Axer, Gerhard Dannecker, Thomas H
Lothar Kuhlen, Eibe H. Riedel, Jochen Tau

Weitere Bände Siehe
http://www.springer.com/series/4333

Peter Dabrock • Jochen Taupitz • Jens Ried
Editors

Trust in Biobanking

Dealing with Ethical, Legal and Social Issues
in an Emerging Field of Biotechnology

 Springer

Series Editors
Professor Dr. Peter Axer
Professor Dr. Gerhard Dannecker
Professor Dr. Dr. h.c. Thomas Hillenkamp
Professor Dr. Lothar Kuhlen
Professor Dr. Eibe Riedel
Professor Dr. Jochen Taupitz (Geschäftsführender Direktor)

Editors

Professor Dr. Peter Dabrock
Dr. Jens Ried
Friedrich-Alexander-University
  Erlangen-Nuremberg
Chair for Systematic Theology / Ethics
Faculty of Philosophy
  and Department of Theology
Kochstraße 6
91056 Erlangen
Germany
peter.dabrock@theologie.uni-erlangen.de
jens.ried@theologie.uni-erlangen.de

Professor Dr. Jochen Taupitz
University of Mannheim
Institute for Medical and Health Law
  and Bioethics
Schloss, Westflügel
68131 Mannheim
Germany
taupitz@jura.uni-mannheim.de

Funded by the German Federal Ministry of Education and Research (grant 01GP0682)

ISSN 1617-1497
ISBN 978-3-540-78844-7   e-ISBN 978-3-540-78845-4
DOI 10.1007/978-3-540-78845-4
Springer Heidelberg Dordrecht London New York

Library of Congress Control Number: 2011940500

Printed on acid-free paper

Springer is part of Springer Science+Business Media (www.springer.com)

Preface

Biobanks are promising instruments of biomedical research and are increasingly considered as essential tools for translational medicine in particular. However, there is concern that the collection of biomarkers in the course of biobanking endeavours could be misused, and thus infringe rights and almost universally accepted ethical standards. In response to these concerns, various sets of governing principles have been established in recent years or are currently discussed in order to protect individuals, families, communities and societies against involuntary use of their data, stigmatisation, discrimination or exclusion that might be caused by data misuse. All efforts addressing these concerns have been grounded on well-established standards of biomedical ethics such as informed consent procedures, protection of individual autonomy, benefit sharing etc. Nevertheless, there are issues that are underrepresented in the ethical, legal and social (ELSI) debates on the challenges posed by biobanks and biobank networks. By highlighting the often neglected aspect of *trust*, this book aims at broadening the horizon of the ELSI-debate and thus filling a gap in current ELSI-research on biobanking.

Apart from being a core issue in the field of ELSI-questions concerning the challenges of biobank research, trust is to be regarded as a focal point for any project relying on biobank infrastructures. Depending on the willingness of potential donors to provide their biospecimen (and additional information) is one of the distinctive features of (at least most non-clinical) biobanks. Therefore, trust in biobanking in general as well as in particular, i.e. in relation to a biobank one considers to contribute to, can assumed to be essential for success and effectiveness of biobank research. Following this basic insight the contributions to this book aim at elucidating meaning, prerequisites and implications of *trust in biobanking*.

This volume contains papers which were presented during two international meetings, held at the Department of Protestant Theology, Philipps-University Marburg, Germany in 2007 and 2008, focussing on ELSI-questions arising in the field of biobank research. Junior researchers from Europe and Canada, representing a broad spectrum of disciplines including ethics, law, philosophy, medicine, social and political sciences and theology, were discussed a variety of issues related to the field of biobanking with international experts. Due to technical reasons, no scientific literature published after 2009 could be incorporated. Nevertheless, we recommend for further reading the opinion "Human biobanks for research" released by the German Ethics Council in 2010 and the Public Health Genomics Special Issue "Privacy, Data Protection, and Responsible Governance. Key Issues and Challenges for Biobanking", edited by Peter Dabrock in 2012.

The first section, *Framing the Field of Biobanking and Trust*, contains basic considerations and, thus, serves as introductory part to the topics this book deals with. In their article "Biobanking: From Epidemiological Research to Population-based Surveillance Systems and Public Health", A. BRAND, T. SCHULTE IN DEN

BÄUMEN and N. PROBST-HENSCH point out how relevant and promising biobank research has proven (or will be proven) to be, not only for medicine (in a more narrow sense), but especially for public health and preventive medicine.

After this introduction to the field from a public health perspective, the following two papers deal with the issue of *trust* from the ELSI-perspective. In "Trust as Basis for Responsibility", C. RICHTER presents a thorough theological and philosophical analysis of *trust*, highlighting social and implications and ethical consequences. K. HOEYER investigates, why measures of trust-building are not only indispensable for any biobank endeavour, but are prerequisites for the effective employment of such a scientific infrastructure. As he argues in "Trading in Cold Blood? Donor Trust in Face of Commercialized Biobank Infrastructures", the fear of commercialization as one of the often mentioned skeptical arguments – especially when private or non-public funded biobanks are discussed – is by far appropriate in any case. Nevertheless, it should not be ignored but seen as a marker pointing to the neglected issue of trust in biobanking.

In the following three sections the ethical, legal and social implications of globalized biobanking are unfolded with special regard to the issue of trust as a necessary prerequisite for successful and effective usage of biobank (infrastructures). The section on *Ethical Issues* is headed by the paper "Which Duty First? An Ethical Scheme on the Conflict between Respect for Autonomy and Common Welfare in Order to Prepare the Moral Grounds for Trust". P. DABROCK goes further into the question, whether or not an obligation to participate in biobank research is defendable and to which extent such an obligation might influence trust-building. C. LENK addresses, based on considerations concerning different interests, the potential role of the traditional principle of justice and fairness for an ethical account of biobank research. His reflections are presented in "Donors and Users of Human Tissue for Research Purposes: Conflict of Interests and Balancing of Interests". The third and closing article of this section is "Collection of Biospecimen Resources for Cancer Research: Ethical Framework and Acceptance from the Patients' Point of View". By assessing an empirical study on demands patients expressed regarding information on and assent to cancer-related biobank research, J. HUBER ET AL. develop a model for specific and need-orientated informed consent procedures.

The third section on *Legal Issues* captures the thread of informed consent which is the core theme of the following papers. Despite the fact that a considerable amount of literature has been published on problematic aspects of informed consent, it is the

S. WALLACE, S. LAZOR and B.M. KNOPPERS provide an overview on existing information and consent materials used by different biobanks, thus introducing the reader to the legal issues of this branch of research. In addition to "What is in a Clause? A Comparison of Clauses from Population Biobank and Disease Biobank Consent Materials", M. SALVATERRA, in "Informed Consent to Collect, Store and Use Human Biological Materials for Research Purposes", suggests a model for a standardized informed consent procedure that regards the needs of potential donors as well as of researchers. The two following articles "Once Given – Forever in a Biobank? Legal Considerations on the Handling of Human Body Materials in Biobanks from a Swiss Perspective" by B. DÖRR and "Biobanks and the Law –

Thoughts on the Protection of Self-Determination with Regards to France and Germany" by K. NITSCHMANN compare and discuss different models of legal regulations in the field of biobanking. As data protection is of special interest for any legal approach to biomedical research in general and biobanking in particular, D. SCHNEIDER elucidates this topic in his paper "Data Protection in Germany: Historical Overview, its Legal Interest and the Brisance of Biobanking".

Finally, S. WALLACE and B.M. KNOPPERS close this section. "The Role of P3G in Encouraging Public Trust in Biobanks" deals with the question, how ethical standards become relevant not only for the communication between science and the general public, but for trust-building, especially when large networks of biobanks are considered.

The last section on *Social Issues* is headed by H. GOTTWEIS' considerations on "Governing Biobank Research", focusing on the political and public challenges posed by emerging networks of biobanks. In "Sharing Orphan Genes: Governing a European-Biobank-Network for the Rare Disease Community", G. LAUSS presents a case-study on the EuroBioBank, investigating how interests of patients might influence research protocols and the development of research infrastructures. Collection, storage and usage of human biological samples is not limited to the western world, but conducted in countries outside Europe and North America. In other cultural contexts, special ethical, legal and social problems might arise, which are not covered by European or US-American standards. The arising challenges concerning this matter are discussed by P. KUMAR PATRA AND M. SLEEBOOM-FAULKNER in their paper "Informed Consent and Benefit Sharing in Genetic Research and Biobanking in India: Some Common Impediments in Practice". Finally, A. GANGULI-MITRA, in "Benefit-sharing, Human Genetic Biobanks and Vulnerable Populations", connects the question on vulnerability as a possible main category for the ELSI-discourse in biobanking with the issue of benefit-sharing, stressing the (often neglected) risk that certain forms of benefit-sharing might intensify existing economic, political, social and cultural inequalities between vulnerable and less vulnerable (parts of the) populations.

The two scientific meetings, taking place in an atmosphere of intense and fruitful discussions, as well as this present book could not have been realized without the help from the whole staff of the Department of Social Ethics at the Faculty of Theology, Philipps-University Marburg, namely Dietmar Becker, Ruth Denkhaus, Elisabeth Krause-Vilmar, Jörg Niesner, Katharina Opalka and Lina Reinartz.

Our special thanks go to Carol George and Dorothee Schönau for her efforts in preparing this publication, again to Jörg Niesner, Katharina Opalka and Lina Reinartz for proof-reading and their considerable help in editing the articles. Last but not least, we owe special thanks to the German Federal Ministry of Education and Research, which funded the two conferences and the publication of this volume (grant 01GP0682). Thankfully, the *Springer Verlag* supported this publication with patience and perseverance.

Erlangen / Marburg / Mannheim 2010

Peter Dabrock
Jochen Taupitz
Jens Ried

Table of Contents

Social Issues

Authors and Editors

Frank Autschbach, MD is Professor of Medicine at Ruprecht-Karls-University Heidelberg (Germany)

Angela Brand, MD, PhD is Professor of Social Medicine and Director of the European Centre for Public Health Genomics at the University of Maastricht (The Netherlands)

Stephan Buse, MD is researcher at the Department of Urology at Ruprecht-Karls-University Heidelberg (Germany)

Peter Dabrock, PhD is Professor of Systematic Theology and Ethics at Friedrich-Alexander-University Erlangen-Nuremberg (Germany)

Bianca Dörr, PhD is Assistant Professor of Law at University of Zurich (Switzerland)

Agomoni Ganguli-Mitra, MSc is Research Fellow at the Institute of Biomedical Ethics at the University of Zurich (Switzerland)

Herbert Gottweis, PhD, is Professor of Political Sciences and member of the Life Science Governance Research Platform at the University of Vienna (Austria)

Esther Herpel, MD is Leader of the Tissue Bank of the National Centre for Tumor Diseases, Heidelberg (Germany)

Klaus Hoeyer, PhD is Associate Professor of Public Health at the University of Copenhagen (Denmark)

Markus Hohenfellner, MD is Professor of Medicine at Ruprecht-Karls-University Heidelberg and Director of the University Hospital for Urology, Heidelberg (Germany)

Johannes Huber, MD, PhD is physician at the University Hospital for Urology, Heidelberg (Germany)

Bartha Maria Knoppers, PhD is Professor of Law and Director of the Centre of Genomics and Policy at McGill University Montreal (Canada)

Georg Lauss, M.A. is researcher at the Department of Political Sciences and member of the Life Science Governance Research Platform at the University of Vienna (Austria)

Stephanie Lazor is Research Assistant at the Centre of Genomics and Policy at McGill University Montreal (Canada)

Christian Lenk, PhD is Assistant Professor of Medical Ethics at Georg-August-University Göttingen (Germany)

Kathrin Nitschmann, PhD is lawyer with specialisation in biomedical law and comparative law

Prasanna Kumar Patra, PhD is Research Fellow at the Department of Anthropology at the University of Sussex (UK)

Nicole Probst-Hensch, PhD is Leader of the Molecular Epidemiology Group at the Centre for Clinical Research at the University of Zurich (Switzerland)

Cornelia Richter, PhD is Assistant Professor of Systematic Theology at Philipps-University Marburg (Germany)

Jens Ried, PhD is Assistant Professor of Systematic Theology and Ethics at Friedrich-Alexander-University Erlangen-Nuremberg (Germany)

Daniel Schneider is lawyer with specialization in health and food law

Tobias Schulte in den Bäumen is Assistant Professor of Law at the University of Maastricht (The Netherlands)

Margaret Sleeboom-Faulkner, PhD is Lecturer at the Department of Anthropology at the University of Sussex (UK)

Jochen Taupitz, PhD is Professor of Law and Director of the Institute for Medical and Health Law and Bioethics at the University of Mannheim (Germany)

Susan Wallace, PhD is Assistant Professor of Law and Head of the Policymaking Core of the International Working Group on Ethics, Governance and Public Participation of the P3G Consortium at McGill University Montreal (Canada)

Framing the Field of Biobanking and Trust

Biobanking for Public Health

Angela Brand, Tobias Schulte in den Bäumen, Nicole M. Probst-Hensch

Abstract Genome-based biobanking requires a new governance model which integrates the personal values of the people concerned, the medical knowledge necessary to define a "genomic indication" as well as the procedural law which enables those professions and families involved to make an ethically and legally acceptable prioritisation of dissenting interests in genomic services and data. Thus, almost all healthcare systems are currently facing fundamental challenges. New ways of organizing these systems based on genomic health information and technologies and stakeholders' different needs are essential to meet these challenges in time.

The issue of biobanking has become a specific challenge having major implications for future research and policy strategies as well as for the healthcare systems in general. The various stakeholders in public health play a key role in translating the implications of genome-based research deriving from biobanks for the benefit of population health. In setting the epidemiological research agenda, in balancing individual and social concerns, by promoting meaningful communication about genomics among researchers, professionals, policymakers, public health agencies, and the public, public health organizations will enhance the potential return on public investment in genomic research. Whereas medicine is currently undergoing remarkable developments from its morphological and phenotype orientation to a molecular and genotype orientation, promoting the importance of prognosis and prediction, the discussion about the role of genome-based biobanking for public health still is at the beginning.

The following chapter contributes to this discussion by focussing on the use of genome-based biobanking for public health research, surveillance systems, health policy development, individual health information management and effective health services.

1 Introduction

The development of target-oriented health promotion, prevention and new treatments in common complex diseases requires the elucidation of the molecular processes involved, the understanding of the causal pathways and the establishment of predictive and diagnostic patterns. To date, epidemiological research and public health practice have been concerned with environmental determinants of health and disease and have paid scant attention to genomic variations within the population as well as between populations. The advances brought about by genomics are changing these perceptions (Peltonen and McKusik 2001, Khoury 1997). Many predict that this knowledge will not only enable clinical interventions but also health promotion messages and disease prevention programmes to be specifically directed and targeted at susceptible individuals as well as subgroups of the population, based on their individual genomic profile and risk stratification. For example, nowadays, it is known that coding variants in DNA determine not only the cause of single-gene disorders, which affect millions of people worldwide, but also predisposition ("susceptibility") (Baird 2000), based on genotype and haplotype variants (Lai et al 2002, Gibbs et al 2002, Probst-Hensch et al 1999), to common complex diseases. The new technologies will allow researchers to rapidly and comprehensively investigate the whole human genome at the level of individual genes (Guttmacher and Collins 2002). Furthermore, there will also be a better understanding of the significance of environmental factors such as chemical agents, nutrition or personal behaviour (Antonovsky 1987) in relation to the causation not only of diseases like osteoporosis, cardiovascular diseases (Sing et al 2003), cerebrovascular diseases, cancer and diabetes, which account for 86% of all deaths and 77% of burden of disease in Europe in 2005, but also of psychiatric disorders, allergies and infectious diseases (Dorman and Mattison 2000, Little 2004, Brand and Brand 2005).

In the past, there had been a narrowed focus looking mainly at the role of inheritance in monogenetic diseases and genetic testing for more than 1000 diseases in that context (human genetics setting). At present, the role of genetic susceptibilities and other biomarkers in common complex diseases is discussed (medical, community health as well as public health setting). In the future, the focus will be even broader by looking at genome-phenomena data sets (Barabasi 2007) and analyzing the role of genomic variants together with other health determinants such as social or environmental factors in health problems (public health setting).

Thus, regarding the understanding of diseases the following "trend" due to novel genome-based knowledge can already be identified (Barbasi 2007, Motter et al 2008, Loscalzo et al 2007, Lunshof et al 2008): recent advances in systems and network biology indicate that specific cellular functions are infrequently carried out by single genes, but rather by groups of cellular components, including genes, proteins, and metabolites. Such a network-based view changes the way of thinking about the impact of mutations and other genomic defects: the damage caused by malfunctioning protein or gene is often not localized, but spreads through the cellular network, leading to a loss of cellular function by incapacitating one or several functional modules ("diseasomes"). New technologies and experimental tools support the systematically mapping of various cellular interactions while enabling to focus not only on the individual components, but also to monitor and analyse the global changes in the cellular network induced by the defective gene or protein. This results in the death of the organism, a finding that may be useful for the design of antibiotics or cancer drugs. Yet for most (genetic) diseases the goal is not to kill the cell, but to recover the lost cellular function or limit the existing damage by asking whether network-based strategies can be developed to predict how to recover function that may have been lost due to defective genes. Research is already starting to change nosology. Seemingly dissimilar diseases are being lumped together. What were thought to be single diseases are being split into separate ailments, i.e. "diseasomes". Just as they once mapped the human genome, scientists are currently trying to map these diseasomes (Barabasi 2007, Motter et al 2008), which can be defined as the collection of all diseases and the genes associated with them. Thus, we are in a unique position in the history of medicine to define human disease precisely, uniquely and unequivocally. It is only a matter of time until these advances will start to affect medical practice as a new field such as "network medicine". The purpose of this perspective is to provide a logical basis for a new approach to classifying human disease that uses conventional reductionism and incorporates the non-reductionist approach of systems biomedicine. What would be the potential of such a systems-based network analysis for the understanding of diseases and their treatments? Loscalzo, Kohane and Barabasi recently identified at least five benefits of the disease network analysis (Loscalzo et al 2007):

1. It can identify those determinants (nodes) or combinations of determinants that strongly influence network behaviour and disease expression or phenotype.
2. It provides unique insight into disease mechanism and potential therapeutic targets.
3. It provides the opportunity to consider the interaction within the network genome, environmental exposures and environmental effects on the posttranslational proteome that define the specific pathophenotype. Thus, disease can be understood as the result of a modular collection of genomic, proteomic, metabolomic and environmental networks that interact to yield the pathophenotype.
4. It provides a mechanistic basis for defining phenotypic differences among individuals with the same disease through consideration of unique genetic and environmental factors that govern intermediate phenotypes contributing to disease expression.
5. It offers a notably method for identifying therapeutic targets or combinations of targets that can alter disease expression.

Overall, the approach offers a novel method for human disease classification, since it defines disease expression on the basis of its molecular and environmental elements in a holistic and fully deterministic way. Although the application of these principles to specific diseases is still in its infancy, the early concepts are internally consistent and the results are encouraging. In addition, the integration of genome-based knowledge into epidemiological and public health research, policies and health services for the benefit of all can be considered as one of the most important future challenges that our health care systems will face (Barbasi 2007, Lunshof et al 2008, Collins et al 2003, Childs and Valle 2000, Collins and McKusick 2001, Burke 2003, Ellsworth and O'Donnel 2004).

Besides that novel biomedical knowledge, also accompanying novel technologies are already triggering the shift in the comprehension of health and disease as well as in the understanding of new approaches to prevention and therapy (Khoury 1996, Brand 2002a, French and Moore 2003). For example, high-throughput and next generation technologies such as tissue microarrays (so-called TMAs) have the potential to screen large numbers of molecular targets in tumor samples for rapid causal, prognostic, diagnostic or therapeutic purposes (Torhorst et al 2001). Complementary to the conventional microarray gene expression profiling, population-based TMAs can be implemented to quickly validate gene expression microarray data in a larger and unselected population of tissue samples (Hoos et al 2001). Through population-based TMAs, it will be possible to assess multiple genomic and protein differences among malignancies such as colorectal cancer, breast cancer, gliomas or rhabdomyosarcoma and thus studying the molecular and cytogenetic changes associated with these malignancies, including human carcinogenic infections (Kononen et al 1998).

Another example of the potential of novel genome-based technologies is the use of nano-chips, which allow the detection of gene activity and genomic pattern by measuring messager RNA (mRNA). By this, it will be possible to predict with higher accuracy and even quicker the response to certain therapies such as interferon therapy.

One of the key questions in all health care systems is whether "the right" interventions and services are provided by the various stakeholders: are the current public health strategies evidence-based? That is, are we assuring the "right" health interventions and innovations (based on combined concepts of health needs assessment and health technology assessment) in the "right" way (based on concepts of quality management and policy impact assessment) in the "right" order and at the "right" time (based on concepts of priority setting and health targets) in the "right" place (based on concepts of integrated health care and health management)?

There has been almost no systematic integration of genome-based knowledge into all of these concepts so far. Current public health strategies are therefore lacking important evidence-based aspects. Thus, with regard to genomics the public health agenda demands a novel vision that reaches beyond the research horizon to arrive at application and public health impact assessment of this novel technology (Brand and Brand 2005, Yoon 2001).

2   The Role of Genome-based Biobanking for Epidemiological Research

The definition of biobanking is very wide and has a twofold character comprising both samples and data. Since biobanks cover therapeutical and population-representative biobanks like blood and tissue banks, including umbilical cord blood banks, semen banks as well as organ collections, they can be defined as collections of samples of human body substances (e.g. cells, tissue, blood, or DNA) that are or can be associated with individual data and information such as clinical, socioeconomic, demographic, lifestyle, behavioural and environmental health determinants.

Biobanking not only allows to store probes of the human body, it also assures the standardisation of sampling processes and data collection. By this, target-orientated preventive, diagnostic and therapeutic interventions can be developed to promote personalized medicine and health care. In the long run, this will result in the provision of more effective and efficient health services.

Furthermore, the already mentioned rapid development of biotechnological research such as population-based TMAs as well as bioinformatics has stimulated the use of biobanks. Although it has been recognized that population-based data on genome-disease and genome-environment interactions are the primary point for assessing the added value of genome-based information for all health interventions in the different health care settings, this approach is not new at all. Human body substances of all kinds have been collected, stored and used for a variety of purposes for many years. Large epidemiologic cohort studies such as EPIC (European Prospective Investigation into Cancer and Nutrition), ARIC (Atherosclerosis Risk in Communities) (ARIC 1989), ALSPAC (Avon Longitudinal Study of Parents And Children), ISAAC (International Study of Asthma and Allergies in Childhood), EUROCAT (European Surveillance of Congenital Anomalies) or various cancer registries and neonatal screening programmes (e.g., in Denmark or Western Australia) have already been able to perform genotyping to expand their existing databases for studying disease incidence and prevalence, natural history and risk factors (Peltonen and McKusick 2001). In addition, large cohort studies such as in the UK (Wright et al 2002) or even involving whole populations such as in Iccland or Estonia (Hakonarson et al 2003) have been initiated to establish repositories of biological materials for the study and characterization of genomic variants associated with common diseases. These biobanks will allow quantifying the occurrence of diseases and risk patterns in various populations and subpopulations as well as to understand their natural histories and risk factors, including genome-environment interactions (Khoury, Little and Burke 2004, Khoury et al 2004).

Nevertheless, the majority of existing biobanks are still relatively small collections of tissue samples related to specific diseases such as cancer. They have been established, for example, in university departments (e.g., in clinics for pathology) or in cancer registries and contain a few hundred up to a few thousand human biospecimen. These biobanks will remain important in the future. But in addition large-scale population-based biobanks have to be established enabling research not only to study single diseases but also "diseasomes" based on individual genome-phenomena data sets (Loscalzo et al 2007, Lunshof et al 2008) and also approaches to a wide range of other health-related issues.

In most countries, besides poor access to human biospecimen, one major bottleneck for large scale biomedical research is the fragmentation of biorepositories. Biobanks may be organized in different clinical settings, in the public sector or in pharmaceutical companies. Irrespective of the responsible institution for biobanking, they may be funded from public or private resources and they may also have been established and used to serve a variety of interests – for instance, purely scientific interests, the interests of donors or commercial interests.

In addition, for most common complex diseases, the collection of body samples for genome-based association studies has often been retrospective in nature and has also been limited to cases of a particularly pronounced phenotype, or with a strong family history. But in order to be able to evaluate the relative risk of a given genomic variant retrospectively from case-control studies, its background frequency in the sample population must be known. Thus, there is a need to recruit large samples of unselected controls from the populations of interest as well as to extend the common cross-sectional or retrospective ascertainment of phenotypes for the prospective follow-up of at least a subset of cases, defined for example by an incidence cohort.

This means that on the one hand long running cohort studies – starting as early as possible in life and including nested case-control studies at various ages and at various occasions – have to be established. This will be a costly, long-lasting, but nevertheless essential public health task. On the other hand another – less costly and less time-consuming – public health task could be the implementation of case-control studies in the very old population to generate hypotheses on genomic-environmental associations, on epigenomic effects as well as on pleiotropic effects.

One specific biobank which has not often been recognized in most countries as an already existing nationwide genome-based biobank is the newborn screening. It has been established for decades in the public sector, in private hand or in public private partnership. Recently, not only the possibility of reanalysing up to 25-year-old Guthrie cards has been discussed. There will be a discussion about shifting from newborn screening exclusively on metabolic diseases to a DNA-based newborn screening for genomic variants as well. A major point of societal discussion will be the question for which validated genomic variants, in addition to metabolic diseases, should newborns be tested for in the future. Should they be screened for complex diseases with the highest burden of disease (e.g., for cardiovascular diseases, cerebrovascular diseases, diabetes, cancer and osteoporosis accounting for 77% of burden of disease in 2005 in Europe), or for orphan diseases (accounting for 10% of all diseases in the whole population and having involved highly validated genomic variants) by developing a resequencing chip for orphan diseases?

Based on these needs the future challenges for biobanks with respect to epidemiological research and public health are comprehensive as well as manifold. They include the promotion of public private partnerships, the linkage of records (e.g., perinatal quality assurance programs, hospital discharge data, data from registries) with data from (genome-based) samples in addition to population-based (mega)biobanks. They also include the integration of genome-based information into the many already existing population-based surveillance systems such as into surveillance systems for infectious diseases, congenital malformations or even into health observatories and the integration of genome-based knowledge into future surveillance systems covering health problems and linking individual information during the whole lifespan (record-linkage based surveillance). Thus, genome-based biobanks can be used as a basis for individual genomic profiling as well as a tool for individual health information management.

Population-based data on genome-disease association and genome-environment interaction form the basis for studying the added value of genome-based health information in various health care settings. They will help to better understand the contribution of genomic variants to common diseases. In the meantime, there is a need to consider how best to collect and monitor information stemming from genome-based research and technologies, to close gaps and to frame the policy development of evidence-based strategies in that field. Thus, the argument of biobanking in the context of public health surveillance systems seems to be crucial.

3 From Epidemiological Research to Population-based Surveillance Systems and Public Health

So far, biobanking and surveillance systems have been looked at independently from each other. This is astonishing, since in the last decades, the concept of surveillance has been quite successfully developed and implemented in various fields of public health. Considerations of problems like data protection or data sharing and also the development of processes and methods in the context of surveillance programs could be easily translated to biobanks.

The idea of observing, recording and collecting facts, analyzing them and considering reasonable health interventions is very old and stems already from Hippocrates (Eylenbosch and Noah 1988). However, before a large-scale organized system of surveillance can be developed, certain requirements need to be fulfilled such as an organized health-care system, a classification system for diseases as well as appropriate methods of measurements. Currently, surveillance is defined as the ongoing systematic collection, analysis, and interpretation of outcome-specific data for use in planning, implementation and evaluation of public health practice (Langmuir 1963). It includes the functional capacity for data collection and analysis as well as the timely dissemination of these data to persons who can undertake effective prevention and therapy. Surveillance data tell where the problems are, who is affected and where effective and efficient health interventions should be directed. Such data can also be used for defining public health priorities in a quantitative manner and to evaluate the effectiveness of programmes. Furthermore, the analysis of surveillance data enables researchers, especially epidemiologists, to identify areas of interest for further investigation.

The uses of surveillance systems are numerous involving quantitative estimates of the magnitude of health problems in a population at risk, analyzing the natural history of diseases, assessing differences by geographic areas, detecting and documenting the spread of health events, identifying research needs to facilitate epidemiologic and laboratory research, testing hypotheses about the etiology of diseases, identifying differences in health status within racial or other subgroups of the population, evaluating health interventions such as preventive or curative strategies, monitoring changes in the nature of diseases, long-term trends or changes in health practices as well as fostering strategic health planning.

Especially the use of registries for surveillance and other medical or public health interventions has increased in the last few years. Registers such as cancer registries differ from other sources of surveillance data in that information from multiple sources is linked for each individual over time. Information is collected systematically from diverse sources including hospital-discharge data, treatment records, pathology reports and death certificates. This specific type of registry is also suitable to monitor health events in groups with increased exposure to hazardous agents. Nevertheless, population-based registries are particularly useful for surveillance because, using incidence rates, the occurrence of a health event can be estimated over time in different geographic areas and subgroups of the population.

The availability and value of data for surveillance depend on a number of factors. These factors include the extent to which classification schemes are used to categorize diagnosis, signs, symptoms, procedures, and reasons for health care, the extent to which information for individuals from different administrative sources over time periods can be linked using a unique personal identifier such as in Denmark or Western Australia. Here, the integration of genome-based biobanks and technologies such as TMAs or nano-chips will be a specific challenge.

In the future, several developments are expected to contribute to the evolution of surveillance systems such as the implementation of bioinformatics, the ability to make more effective use of sophisticated epidemiological and statistical tools to detect changes in patterns of health problems, the electronic dissemination of surveillance data and – last but not least – novel knowledge and innovations such as genome-based knowledge and technologies. The critical challenge, however, is the need to regard surveillance as a scientific enterprise. To do this properly, the principles of surveillance and their role in guiding epidemiologic research and in influencing other aspects of the overall mission of public health have to be fully understood. In addition, new epidemiological methods based on public health surveillance have to be developed. Bioinformatics for efficient data collection, analysis and dissemination have to be applied. Ethical, legal and social concerns have to be addressed right in time, the benefit of surveillance systems has to be reassessed on a routine basis, and surveillance practice has to be translated into emerging areas of public health practice such as the integration of genome-based biobanks.

The success of these surveillance systems including genome-based biobanks will heavily depend on the quality of the information into the system (i.e., on validated population-based genomic variants) and on the value of the information to its intended users. A clear understanding of how policymakers, voluntary and professional groups, researchers, the commercial sector and other stakeholders might use surveillance data is valuable in gathering the support of these audiences for the surveillance system.

Regarding data sharing, it has to be stressed that different sources of information need to be accessed and compared with or added to the data collected in its own system, e.g., laboratory results, tissue results, epidemiological information for specific conditions, population estimates and mortality records. Through responsible planning and coordination on the part of managers on reporting systems, standard coding schemes can be adopted as data systems evolve. These actions, for example, have the potential to facilitate the sharing and use of data.

European and US public health institutions and platforms like the Public Health Genetics Foundation in Cambridge (PHGF), UK, the European Centre for Public Health Genomics in Maastricht, the Netherlands, the Turkish Center for Public Health Genomics and Personalized Medicine (TOGEN) in Ankara or the US National Office of Public Health Genomics (NOPHG) at the Centers for Disease Control and Prevention in Atlanta (CDC), who work closely together with researchers from genetic and molecular science ("modern biology") as well as from population science, humanities and social science, are optimistic and clear about the relevance of the integration of genome-based biobanks into surveillance systems and thus, for public health in general (Brand et al 2004, Khoury et al 2000, Omenn 2000, Walt 1994). Interestingly enough, they all have strong links or are even part of the respective national genome research projects in these countries and are translating genome-based knowledge from biotechnology and biobanks through genetic epidemiology or "classical" epidemiology into public health ("translational research"). By using methods like horizon scanning, fact finding and monitoring to identify research trends as early as possible, they are already doing a prospective evidence-based evaluation. That is an evaluation that is already carried out in the process of basic research and not just in the (retrospective) process of the implementation of public health strategies and policies (Williams 2005), which always will tend to lack behind.

4 Public Health Ethics as a New Paradigm

The present discussion about ethical aspects of biobanking is dominated by the conventional and individual-centered moral categories of medical ethics and bioethics. But especially in this context of biobanking, focussing always on individual rights and protections such as informed consent, confidentiality, discrimination, stigmatization or the "right not to know" in the end will undermine individual rights and interests in ways that benefit some organized interests, because important social, political and scientific questions will be hardly considered (Schröder 2007).

Regardless of the question of whether in the situation of "informed consent" the promises and information that potential research subjects are given are accurate indeed, and regardless of the point that informed consent is irrelevant to many groups of potential research subjects, the biggest concern is that by focussing on individual consent to research, the importance of statutory research like for example epidemiological and public health research or the monitoring of the effectiveness and efficiency of specific health interventions, that poses few risks to individuals but is essential to the assurance and improvement of (collective) health care provision, is neglected. This focus on individual decision-making not only ignores contexts of choice but also is connected to a view of ethics that is separated from politics.

Moreover, if ethics in the context of biobanking is further promoting informed consent and confidentiality, then ethics is no longer concerned with the scientific validity of research, balancing likely benefits of research and establishing research priorities. Thus, the question of whose health and whose interests will be served by research is critical. Public policy principles such as institutional oversight and competing political priorities, ethical principles such as solidarity, justice and good governance as well as the concepts of benefit-sharing and informed contract will be able to provide an appropriate public health ethics framework (Schröder 2004, Sass 2001, Knoppers 2005). But continuing to make an artificial division between ethical and political aspects on the one hand and between individual rights and public goods on the other hand, will impede any innovative future use of biobanking.

Since policy development in the field of biobanking must take contextual as well as cultural factors into consideration, "both collective and individual rights and interests are at stake in creating or assessing genomic databases for public health research" (Coughlin 2006). At the same time, a public health ethics framework (Knoppers 2005)[1] must be based on norms beyond the legal and ethical criteria of autonomy and privacy. Furthermore, new models are needed to offer robust moral guidance while keeping the reality of a dynamic science such as the concept of diseasomes in mind (Loscalzo et al 2007).

Also, since biobanks require an ongoing contribution from the potential research subjects, if not samples, then at least health information and possible lifestyle data need to be collected over an indefinite period of time. It becomes obvious, that in this situation complete anonymisation of data is impossible, as this would prevent new data being linked to the old and to tissue samples or genome-based information.

Clarifying the general conditions under which genome-based knowledge and technologies can be put to best practise in the field of biobanking, epidemiological research and public health surveillance, paying particular consideration to the public health specific ethical, legal and social implications (ELSI) (Brand et al 2004, Omenn 2000, Michigan Center for Genomics and Public Health 2004), is currently the most pressing task and thus has been stressed by public health genomics (PHG). Aiming the application of genetic and molecular science to the promotion

---

[1] Cf. http://www.ete-online.com/content/3/1/16, Accessed 15 July 2008.

of health and disease prevention through the organised efforts of society, integral to its activities is dialogue with all stakeholders in society, including industry, governments, health professionals and the general public (Walt 1994).

Policymakers must be aware of the current challenge to improve consumer protection, to monitor the implications of genome-based knowledge and technologies for health, social and environmental policy goals and to assure that genomic advances will be tailored not only to treat medical conditions, but also to prevent disease and improve health (Beskow et al 2001). Sound and well reflected genetics policies and programs require a timely and coordinated process for evidence-based policy making that relies on scientific research and ongoing community consultation (Frankish et al 2002). An acceptable and maybe delicate balance between providing strong protection of individuals' interests (O Neill 2002, Geier and Schröder 2003) and enabling society to benefit from the genomic advancements at the same time must be found (Brand et al 2004, Beskow et al 2001, Tauber 2003, UNESCO 2003).

Interestingly enough, research in biobank-related fields such as for example in the field of genetically modified food has been able to distinguish between trust in governance, trust in government and trust in non-governmental organizations (NGOs), since trust is an important predictor of the attitude of individuals towards innovations, public acceptance, technological optimism, and various forms of behaviour. Public trust as a rather complex social and political phenomenon remains an important issue in the near future, especially in the context of the integration of genome-based biobanks into public health surveillance systems. It is not just the restoring of public faith in government, industry, NGOs or other stakeholders. It is related to the way government or politicians are willing to involve the public within decision-making, how industry is handling consumer interests and individuals' perception of the way biotechnology may influence their lives (Hansen et al 2001). With higher levels of trust in governance, even more important than trust in government or NGOs, people have a more positive attitude towards an innovation, are more likely to accept new knowledge and technologies, and are more optimistic about technological developments (Gutteling et al 2006). E.g., also regarding the implementation of population-based biobanks policy-makers and clinicians should consider how to narrow the gap between expectations and reality (Coulter and Jenkinson 2005).

Since the effectiveness and efficiency of biobanks depend very much on people's willingness to contribute samples for both research and storage, public support is thus essential in assuring long-term realisation and potential of biobanks. It is also based on the assumption that the complex issues surrounding biobanks are managed appropriately by the responsible authorities. Although the majority of the general public is willing to donate a sample to a biobank (Kettis-Lindblad et al 2005), the willingness is mainly driven by altruism and depends on the public well-informed and having trust in experts and institutions. In general, there is an overwhelming positive attitude towards genomic research. Nevertheless, the trust in authorities' capability to evaluate the chances and risks of genomic research varies, whereas individual university/hospital-based researchers receive the greatest trust, while health care providers and the politicians receive the lowest trust. Most individuals (86%) would donate a linked blood sample for research purposes and would also agree to both donation and storage (78%). Besides that, the most common motive is the benefit of future patients (89%) as well as for the benefit of themselves or their families (61%). Those more likely to donate a sample are middle-aged and have children, which may be explained by the theory about generativity (Erikson and Kivnick 1986), have a genetic disease in the family or among close friends, are blood donors, have a positive attitude towards genomic research and have high trust in experts and institutions. Only 1% refuse to allow their tissue to be used for commercial research. Consequently, maintaining or improving the public's trust seems to be as important as having an informed public.

Independent from the ethical points mentioned above, it should be allowed to argue that at present "reinventing the wheel" seems to be very true in the discussion around biobanks. Almost all ethical and legal aspects which have been discussed so far in detail in the context of biobanking, especially in the context of genome-based biobanking, are neither new nor exceptional. Already for several years there has been a rich and growing body of literature on ethical issues in epidemiological research and public health practice including conceptual frameworks of public health ethics. Attention has been given to issues such as generalizable knowledge by elucidating the causes of disease, by combining epidemiological data with information from other disciplines such as genomics and microbiology, by evaluating the consistency of epidemiologic data with etiological hypotheses and by providing the basis for developing and evaluating health promotion and prevention procedures (Seigel 2003). Ethics guidelines have been developed, e.g. for the Industrial Epidemiology Forum, the International Society for Environmental Epidemiology, and the American College of Epidemiology. The latter one discusses core values, duties and virtues in epidemiology, the professional role of epidemiologists, minimizing risks and protecting welfare of research participants, providing benefits, ensuring an equitable distribution of risks and benefits, protecting confidentiality and privacy, obtaining informed consent, submitting proposed studies for ethical review, maintaining public trust, avoiding conflicts of interest and partiality, communicating ethical requirements, confronting unacceptable conduct and obligations to communities.

Thus, benefit of public health surveillance must be balanced against possible risks and harms, such as infringements on personal privacy. There is also the need to balance health as a value with values of privacy and autonomy. Above all, there is a need for sensitivity ethnic and cultural habits and norms. Such concerns have been addressed through participatory community-based research.

The interest of ethical issues in epidemiological research and public health reflects both the important societal role of public health and the growing public interest in the scientific integrity of health information as well as the fair distribution of health care resources (Seigel 2003, Gostin 2001).

5 Public Health Law as a New Paradigm

The sharing and linkage of data and samples has been identified as a key to the long-term success of biobanks and genomic research. Studies of the European Commission have revealed uncertainty with regard to the harmonisation in Europe (European Science Foundation 2008). The existing directives and regulatory documents of the EC and its agencies cover certain areas but not all aspects of biobanking. Due to the competences of the EC and its Member States the field of data protection is fully harmonised while the collection, storage and sharing of samples is not. As researchers are not interested in the biological material as such, but in the data contained, this approach seems artificial to the research community. The inconsistent governance encumbers the progress in biobanking and pharmacogenomics and steps should be taken to overcome the hurdles. From a conceptual perspective data protection law serves the idea of an information freedom. Thus, research with health data is privileged in data protection if certain conditions are met. Data protection could be used as the backbone of an unified governance for the sharing and linking of data and samples in Europe if the Data Protection Directive is interpreted in a harmonised way by authorities and, in particular, by ethics committees.

The linkage and sharing of data and samples is legal if biobanks and researchers meet the legal requirements set by data protection law and the regulations which govern the collection and use of samples. While the data protection law is fully harmonised, a unified governance system for samples is still missing. As the scientific and economic value of biobanks is determined by the medical and secondary data stored in the biobank the overall governance in public health should be based on data protection principles. Data protection comprehends vehicles which enable stakeholders to balance conflicting rights. Thus, data protection can be used in a positive way to reach a standard of information justice. The current problems with the application of law seem to derive from the uncertainty as to how the regulations need to be interpreted. With regard to the acting forces in the field it could be either the research community which, in a sort of anticipatory obedience, does not explore the potential of biobanking, or the ethics committees which are setting up higher standards than the ones foreseen by law. From a legal perspective only very few issues still need to be solved. The biggest concern is the lack of purpose specificity of large-scale biobanks. The problems deriving from this could be overcome by an optimised proband-oriented disclosure and by technical means and concepts (like coding and trustee models). On a more conceptual level fears related to the sharing and linking of data could also be reduced if the research secrecy is better protected. A research clause which is not profession-, but disease-outcome oriented, could be one task for a further harmonisation of data protection law in Europe. An assessment is also necessary with regard to Art 8 para 4 and Recital 34 of the Directive, as this option enables Member States to set up different data protection levels for sharing and linkage. Otherwise the harmonised interpretation and the development of an overall governance scheme of biobanking are essential for the future exploitation of biobanks in Europe. This governance framework in public health should be research friendly and non-discriminatory.

6 Conclusion

Neither any single large cohort study nor any other single epidemiological population-based study will have adequate statistical power to examine all potentially relevant genomic variants, their interactions with each other and with environmental factors, gene expression profiles and proteomic patterns. Thus, national and international collaboration is essential to realize the full potential of biobanks and other large-scale population studies and also to develop and apply standardized epidemiological methods for assessing genomic variants in populations by means of systematic reviews, training and dissemination of information.

New genome-based information and technologies will force health communities to enhance surveillance systems by integrating this knowledge arriving from biobanks as well as to enhance epidemiologic capacity for collecting and analyzing information stemming from community-based assessments of genomic variation (Annas 2000), providing evidence about the burdens of various diseases. As with other fast-paced scientific and technological advancements, the intersection between genomics and public policy will continue to require close monitoring using public health methods like health technology assessment (HTA) (Banta and Luce 1993, Pollit et al 1997, Brand 2002b, Moldrup 2002, Perleth 2003, AETMIS 2003), health needs assessment as well as health impact assessment and will also continue to require timely action. By this, there will be a chance to ensure the appropriate and responsible use of genome-based information and new technologies (Shani 2000).

The various stakeholders in public health play a key role in translating the implications of genome-based research arriving from biobanks for the benefit of population health. In setting the epidemiological research agenda, in balancing individual and social concerns, by promoting meaningful communication about genomics among researchers, professionals, public health agencies, and the public, "... public health organizations will enhance the potential return on public investment in genomic research" (Gwinn and Khoury 2006).

Policymakers now have the opportunity to protect consumers, to monitor the implications of genomics for health services and to assure that genomic advances will be taped to prevent disease and improve health by analysing

- the history of biobanks and the purposes they were set up for,
- the translational process from basic knowledge generated in population-based biobanks to the development of public health policies, interventions and programmes,

- the timely and responsible integration of genome-based health information and technologies into public health research, policy and practice in the different healthcare systems,
- the ability of biobanks to serve researchers and other relevant stakeholders with a particular public health perspective,
- the design of biobanks and their preparedness to provide data on evidence-based risk management and
- the place of biobanks in legislations related to public health.

This will be a doable project (Smith et al 2005), but will require regional as well as European as well as global coordination (Daar 2002). The next decade will provide a window of opportunity to establish infrastructures, across Europe and globally, that will enable the scientific advances to be effectively and efficiently translated into evidence-based policies and interventions that improve population health (Brand 2005).

Acknowledgments   This chapter is a result of the work of the Public Health Genomics European Network (PHGEN, www.phgen.eu) (Bosch 2006), which is funded in the Public Health Programme of the European Commission (Project Number 2005313). Nevertheless, the views described reflect the authors' views.

References

AETMIS (Agence d'evaluation des technologies et des modes d'intervention en sante) (2003) Health Technology Assessment in Genetics and Policy-making in Canada : Towards a sustainable development. Report from a symposium held September 11th and 12th, 2003 in Montreal,

Annas GJ (2000) Rules for Research on Human Genetic Variation – Lessons from Iceland. N Engl J Med 342: 1830-1833

Antonovsky A (1987) Unraveling the mystery of health. How people manage stress and stay well. Jossey Bass, San Francisco

ARIC investigators (1989) The Atherosclerosis Risk in Communities (ARIC) Study: Design and objectives. Am J Epidemiol 129: 687-702

Banta HD, Luce BR (1993) Health Care Technology and its Assessment. An International Perspective. Oxford University Press, Oxford et al

Baird PA (2000) Identification of genetic susceptibility to common diseases: the case for regulation. Perspect Biol Med 45: 516-528

Barabasi AL (2007) Network Medicine – From Obesity to the "Diseasome". N Engl J Med 357: 1866-1868

Beskow LM et al (2001) The Integration of Genomics into Public Health Research, Policy and Practice in the United States. Community Genet 4: 2-11

Bosch X (2006) Group Ponders Genomics and Public Health. JAMA 295: 1762

Brand A (2002a) Prädiktive Gentests – Paradigmenwechsel für Prävention und Gesundheitsversorgung? Gesundheitswesen 64: 224-229

Brand A (2002b) Health Technology Assessment als Basis einer Prioritätensetzung. In Fozouni B, Güntert B (eds) Prioritätensetzung im deutschen Gesundheitswesen. Logos Verlag, Berlin

Brand A (2005) Public health and genetics – dangerous combination? View-point section. Eur J Public Health 15:113-116

Brand A, Brand H (2005) Public Health Genetics – Challenging "Public Health at the Crossroads". Ital J Public Health 2: 59

Brand A, Dabrock P, Paul N et al (2004) Gesundheitssicherung im Zeitalter der Genomforschung – Diskussion, Aktivitäten und Institutionalisierung von Public Health Genetics in Deutschland. Gutachten zur Bio- und Gentechnologie. Friedrich-Ebert-Stiftung, Berlin

Burke W (2003) Genomics as a Probe for Disease Biology. N Engl J Med 349: 969-974

Childs B, Valle D (2000) Genetics, Biology and Disease. Ann Rev Gen Hum Genet 1: 1-19

Collins FS, McKusick VA (2001) Implications of the Human Genome Project for Medical Science. J Am Med Ass 285: 540-544

Collins FS, Patrinos A, Jordan E et al (1998) New Goals for the U.S. Genome Project: 1998-2003. Science 282: 682-689

Coughlin S (2006) Ethical issues in epidemiologic research and public health practice. BioMed Central Open Access. Emerging themes in Epidemiology 3

Coulter A, Jenkinson C (2005) European patients' views on the responsiveness of health systems and healthcare providers. Eur J Public Health 15: 355-360

Daar AS (2002) Top 10 biotechnologies for improving health in developing countries. Nat Genet 32: 229-232

Dorman JS, Mattison DR (2000) Epidemiology, Molecular Biology and Public Health. In Khoury MJ, Burke W, Thomson EJ (eds.) Genetics and Public Health in the 21st Century. Using Genetic Information to Improve Health and Prevent Disease. Oxford University Press, Oxford et al

Ellsworth DL, O'Donnell CJ (2004) Emerging Genomic Technologies and Analytic Methods for Population- and Clinic-Based Research. In Khoury MJ, Little J, Burke W (eds) Human Genome Epidemiology. A Scientific Foundation for Using Genetic Information to Improve Health and Prevent Disease. Oxford University Press, Oxford et al

Erikson JM, Kivnick HQ (1986) Vital involvement in old ages. Norton, New York

European Science Foundation (2008) Population Surveys and Biobanking. Science Policy Briefing. www.esf.org. Accessed 20 May 2008

Eylenbosch WJ, Noah ND (1988) Historical aspects. In: Eylenbosch WJ, Noah ND (eds) Surveillance in health and disease. Oxford University Press, Oxford

Frankish CJ, Kwan B, Ratner PA et al (2002) Challenges of citizen participation in regional health authorities. Soc Sci Med 54: 1471-1480

French ME, Moore JB (2003) Harnessing Genetics to Prevent Disease and Promote Health. Partnership for Prevention. Washington

Geier M, Schröder P (2003) The Concept of Human Dignity in Biomedical Law. In Sándor J, den Exter AP (eds) Frontiers of European Health Law: A Multidisciplinary Approach. Erasmus University Press-DocVision, Delft

Gibbs RA, Belmont JW, Hardenbol P (2003) The International HapMap Project. Nature 426: 789-796

Gostin LO (2001) Health information: reconciling personal privacy with the public good of human health. Health Care Anal 9: 321-335

Gutteling J, Hanssen L, van der Veer N et al (2006) Trust in governance and acceptance of genetically modified food in the Netherlands. Public Underst Sci 15: 103-112

Guttmacher AE, Collins FS (2002) Genomic Medicine – A Primer. N Engl J Med 347: 1512-1521

Gwinn M, Khoury MJ (2006) Genomics and Public Health in the United States: Signposts on the Translation Highway. Community Genet 9: 21-26

Hakonarson H, Gulcher JR, Stefansson K (2003) decode Genetics, Inc. Pharmacogenomics 4: 209-215

Hansen L, Gutteling JM, Lagerwerf L et al (2001) In the Margins of the Public Debate on "Food and Genes": Research under Commission of the Committee Biotechnology and Food. Twente University, Enschede

Hoos A, Urist MJ, Stojadinovic A et al (2001)Validation of tissue microarrays for immunohisto-chemical profiling of cancer specimens using the example of human fibroplastic tumors. Am J Pathol 158: 1245-1251

Kettis-Lindblad A, Ring L, Viberth E, Hansson MG (2005) Genetic research and donation of tissue samples to biobanks. What do potential sample donors in the Swedish general public think? Eur J Public Health 16: 433-440

Khoury MJ (1996) From Genes to Public Health: The Applications of Genetic Technology in Disease Prevention. Am J Public Health 86: 1717-1722

Khoury MJ (1997) Relationship Between Medical Genetics and Public Health: Changing the Paradigm of Disease Prevention and the Definition of a Genetic Disease. Am J Med Gen 17: 289-291

Khoury MJ, Burke W, Thomson E (2000) Genetics in Public Health: A Framework for the Ingegration of Human Genetics into Public Health Practice. In Khoury MJ, Burke W, Thomson EJ (eds) Genetics and Public Health in the 21st Century. Using Genetic Information to Improve Health and Prevent Disease. Oxford University Press, Oxford et al

Khoury MJ, Little J, Burke W (2004) Human genome epidemiology: scope and strategies. In: Khoury MJ, Little J, Burke W (eds.) Human Genome Epidemiology. Oxford University Press, New York

Khoury MJ, Millikan R, Little J et al (2004) The emergence of epidemiology in the genomics age. Int J Epidemiol 33: 936-944

Knoppers BM (2005) Of genomics and public health: Building public "goods"? Can Med Assoc J 173: 1185-1186

Kononen J, Bubendorf L, Kallioniemi A et al (1998) Tissue microarrays for high-throughput molecular profiling of tumor specimens. Nat Med 4: 844-847

Lai E, Bansal A, Hughes A (2002) Medical applications of haplotype-based SNP maps: learning to walk before we run. Nat Genet 32: 353

Langmuir AD (1963) The surveillance of communicable diseases of national importance. N Engl J Med 268:182-192

Little J (2004) Reporting and Review of the Human Genome Epidemiology Studies. In: Khoury MJ, Little J, Burke W (eds) Human Genome Epidemiology. A Scientific Foundation for Using Genetic Information to Improve Health and Prevent Disease. Oxford University Press, Oxford et al

Loscalzo J, Kohane I, Barabasi AL (2007) Human disease classification in the postgenomic era: A complex systems approach to human pathobiology. Mol Syst Biol 3: 124

Lunshof JE, Chadwick R, Vorhaus DB et al (2008) From genetic privacy to open consent. Nat Rev Genet 9: 406-411

Michigan Center for Genomics and Public Health (2004) Ethical, Legal and Social Issues in Public Health Genetics (PHELSI), http://www.sph.umich.edu/genomics/media/subpage_autogen/PHELSI.pdf. Accessed 15 July 2008

Moldrup C (2002) Medical Technology Assessment of the Ethical, Social, and Legal Implications of Pharmacogenomics. A research Proposal for an Internet Citizen Jury. Int J Technol Assess Health Care 18: 728-732

Motter AE, Gulbahoe N, Almaas E et al (2008) Predicting synthetic rescues in metabolic networks. Mol Syst Biol 4: 168

Omenn GS (2000) Public health genetics: an emerging interdisciplinary field for the postgenomic era. Ann Rev Public Health 21: 1-13

O'Neill O (2002) Autonomy and Trust in Bioethics. Cambridge University Press, Cambridge

Peltonen L, McKusick VA (2001) Genomics and Medicine. Dissecting Human Disease in the Postgenomic Era. Science 291: 1224-1229

Perleth M (2003) Evidenzbasierte Entscheidungsunterstützung im Gesundheitswesen. Konzepte und Methoden der systematischen Bewertung medizinischer Technologien (Health Technology Assessment) in Deutschland. Verlag für Wissenschaft und Kultur, Berlin

Pollitt RJ, Green A, McCabe CJ (1997) Neonatal screening for inborn errors of metabolism: cost, yield and outcome. Boldrewood, The National Coordinating Centre for Health Technology Assessment (NCCHTA)

Probst-Hensch NM, Ingles SA, Diep AT et al (1999) Aromatase and breast cancer susceptibility. Endocr Relat Cancer 6(2):165-173

Sass HM (2001) A 'Contract Model' for Genetic Research and Health Care for Individuals and Families. Eubios J Asian Int Bioeth 11: 130-132

Schröder P (2004) Gendiagnostische Gerechtigkeit. Eine ethische Studie über die Herausforderungen postnataler genetischer Prädiktion. LIT Verlag, Münster et al

Schröder P (2007) Public Health Ethik in Abgrenzung zur Medizinethik. Bundesgesundheitsbl – Gesundheitsforsch – Gesundheitsschutz 50: 103-111

Seigel D (2003) Clinical trials, epidemiology, and public confidence. Stat Med 22: 19-25

Shani S, Siebzehner MI, Luxenburg O (2000) Setting priorities for the adoption of health technologies on a national level – the Israeli experience. Health Policy 54: 169-185

Sing CF, Stengard JH, Kardia SL (2003) Genes, environment, and cardiovascular disease. Art Thromb Vasc Biol 23: 1950

Smith GD, Ebrahim S, Lewis S et al (2005) Genetic epidemiology and public health: hope, hype, and future prospects. Lancet 366: 1484-1498

Tauber AI (2003) Sick Autonomy. Perspect Biol Med 46: 484-495

Torhorst J, Bucher C, Kononen J (2001) Tissue microarrays for rapid linking of molecular changes to clinical endpoints. Am J Pathol 159: 2249-2256

UNESCO (2003) International Declaration on Human Genetic Data. Paris

Walt G (1994) Health Policy: An Introduction to Process and Power. Zed Books, London New Jersey

Williams G (2005) Bioethics and large-scale biobanking: individualistic ethics and collective projects. Genomics Soc Policy 1: 50-66

Wright AF, Carothers AD, Campbell H (2002) Gene-environment interactions – the BioBank UK study. Pharmacogenomics 2: 275-282

Yoon PW (2001) Public health impact of genetic tests at the end of the 20th century. Genet Med 3: 405-410

Zimmern R, Cook C (2000) Genetics and Health. Policy issues for genetic science and their implications for health and health services. The Nuffield Trust Genetics Scenario Project. The Nuffield Trust, London

Trading in Cold Blood?

Trustworthiness in Face of Commercialized Biobank Infrastructures

Klaus Hoeyer

Abstract In this chapter I discuss trustworthiness as a quality of biobanks constructed and regulated in ways that make them serve the expectations donors have. Beginning with a short review of empirical studies of donor expectations and attitudes, I show that their concerns tend to revolve around issues of 1) personal control, 2) harmful uses of medical knowledge and 3) social fairness, in particular whether research is shaped by creed rather than medical need. I then rehearse the regulatory tendencies addressing these three issues, and identify an overall trend towards commercialization in particular in the areas of property law and research management. Then, the potential effects of this commercialization on the research agenda and research results are assessed based on available empirical studies. In conclusion, I point to the gap between donor expectations and the thrust of regulatory efforts. I argue that it is important that ethicists begin to address the wider innovation system surrounding biobanks if they wish that biobanks do not only preserve, but also deserve, the trust of the donating public.

1 Introduction

Public trust is a prerequisite for the viability of population-based biobanks. As such, it can be considered a valuable asset for the researchers and authorities responsible for the construction of large-scale biobanks. From the perspective of the research participant, however, trust in itself is not necessarily valuable; donors are typically more interested in the trustworthiness of the institutions handling their biological material and data. If trustworthiness in this context means that biobanks are capable of and committed to pursuing the goals that donors expect them to pursue, this chapter looks at the relationship between trustworthiness and research regulation: what do we know about donors' concerns and expectations, on the one hand, and about regulatory trends and their implications for medical research, on the other hand.

Legislation that deals explicitly with biobank governance has mainly focused on informed consent – a requirement that is occasionally described as pivotal in preserving donor trust (Clayton 2005, Sutrop 2007). In this chapter I claim that biobank governance must be understood in a wider context of innovation politics, including regulatory measures such as property law, research management and funding regimes. Current innovation politics stimulate what I call 'commercialization', a process that has potential for contradicting the interests expressed by donors. I argue that debates about ethics have been too narrowly focused on the level of information that are to be provided to donors and that donors' real concerns, and thus the trustworthiness of biobank governance, require scrutiny of the overall regulatory frameworks for biobank research. The fact that people have signed a consent form, the content of which they might not have understood or remembered, is likely to be of little help if people gradually discover that their expectations are not met. If biobanks are not only to establish or preserve, but also deserve public trust, we must address the wider innovation system in which biobanks operate.

The chapter begins with a short rehearsal of the main insights generated by studies of the expectations and concerns of donors. The following section explores the policy areas that affect the ability to meet these expectations and concerns. It demonstrates that many features of the regulation of innovation point towards a commercialization process in the medical research field. The general implications of this tendency are assessed in the next section of the paper. It is argued that commercialization might effect negative changes of a type that concerns donors: decision-making that is increasingly based on assessments of commercial rather than medical need. Finally, in the conclusion, I point to the need to refocus biobank ethics to this wider set of issues and to move ethical debate beyond the consent issue.

A chapter seeking to deliver a broad outline of donor concerns, regulatory measures and implications of commercialization naturally cannot claim to exhaust the issues. Rather than attempting to deliver a comprehensive presentation, the chapter wishes to point to some important but under-explored regulatory contexts in urgent need of attention for the sake of continued donor trust.

2 What Are Donors Concerned About?

Qualitative empirical studies of donor expectations and concerns indicate that the majority of donors are positive, even if ambivalently so, towards medical research and its implications (Busby 2004, Weldon 2007, Hoeyer 2004). Although they are largely inclined to support medical research, they have concurrent worries about potential misuses and unwarranted effects. Studies have highlighted how people in the European welfare states see participation as a type of obligation connected with the benefits of universal healthcare (Busby 2004, Busby 2006, Hoeyer 2003, Skolbekken et al 2005, Svendsen 2007). In observation and interview studies the information provided to donors prior to research participation has been shown to have a limited effect on their willingness to participate (Busby 2006, Hoeyer 2003, Ducournau 2007). In fact, donors rarely read, recall or use the information sheet. People are motivated by the context in which they are invited to participate, rather than by the information delivered – and they expect medical research facilitated by public health care institutions to benefit everybody (Ibid.).

Despite a generally positive attitude towards medical research, the published qualitative studies indicate that three areas of concern are common among the donors:

1. Donors reflect on the lack of control they experience once tissue has left their body (Levitt and Weldon 2005). They express awareness of the fact that they essentially have to trust the research institutions as they have no opportunity for subsequent inspection (Hoeyer 2003).
2. Donors express concern about potentially harmful uses of medical research. They focus in particular on confidentiality issues (insurance companies or employers getting hold of genetic information etc.), the use of genetic knowledge for cloning or selective abortions, enhancement technologies and stigma. There is also a concern that researchers pursue their curiosity without due caution.
3. Finally, most donors contemplate issues of social fairness. A particular worry is that creed rather than need determines the research agenda or that there will not be equal access to research results (Bister et al forthcoming, Barr 2006, Busby 2006, Ducournau 2007, Haimes and Whong-Barr 2004, Skolbekken et al 2005, Haddow et al 2007, Hoeyer 2006). The issue of social fairness is apparently particularly important, as donors not only express what could make them abstain from participation, but indicate their motivation on positive terms. For example, they volunteer to help producing new treatments in the expectation that the developed treatments are really needed and will be affordable for everyone.

Quantitative surveys have also been conducted to measure the attitudes among donors towards the use of human biological material for research, even if they provide a more heterogeneous picture than the qualitative studies of donor reasoning. Despite the heterogeneity of results, it is possible to infer a few general insights from the survey studies:

1. It seems that the type of tissue requested and the position of the donors in relation to the research project influence the view of tissue research. Cancer patients are generally very supportive of research on their tissue (Malone et al 2002, Pentz et al 1999) and potential participants in cohort studies are less interested but still relatively supportive (Cousins et al 2005, Kettis-Lindblad et al 2006, Kettis-Lindblad et al 2007), while the kin of potential cadaveric donors are least likely to accept tissue donations (Womack and Jack 2003, Womack et al 2006). The social groups most likely to abstain vary depending on the national context: for example, in the US ethnic minorities and people with poor education will typically abstain (Pentz et al 2006, Wendler and Emanuel 2002),

while in Sweden it seems to be younger men with higher education who are most likely to abstain (Hoeyer et al 2004, Kettis-Lindblad et al 2007).
2. Most donors think that they should have a say concerning the retention of tissue (e.g. Nilstun and Hermerén 2006, Stegmayr and Asplund 2002, Wendler and Emanuel 2002). This is typically interpreted to be in support of a consent requirement (Merz 1997), but whether it is indicative of support for broad (Wendler 2006) or specific consent (Shickle 2006) varies remarkably in the literature reviews.
3. Though many donors accept commercial access to public biobanks (Jack and Woomack 2003, Stegmayr and Asplund 2002), it seems to be viewed more as a necessary evil than as the preferred research infrastructure (e.g. Cragg et al 2002, Gudmundsdóttir and Nordal 2007). In Canada, a majority accepts DNA patenting but expresses worries about issues of access and affordability (Earnscliff Research and Communications 2000).

There always seems to be at least one survey contradicting every finding. Nevertheless, we might tentatively conclude that 1) donors, to become donors, must trust the institution; 2) donors expect their wishes regarding the use of their tissue to be respected; and that 3) the trust of donors is potentially compromised by commercialization, unless it is handled in a manner that ensures attention to the objectives lauded by donors.

Although donors generally express nuanced reasoning about research and the regulation of research, they rarely see themselves as engaging in the regulatory process through the consent procedure. Instead, they expect the authorities to take responsibility for addressing their concerns (Hoeyer 2003), in particular, the extent to which the research agenda addresses public health needs.

3 What Regulates the Issues that Donors are Concerned About?

If donors contemplate their own lack of control, and express concerns about issues of harmful use of medical knowledge and the absence of social fairness – what, then, is the regulation of medical innovation doing to address these issues?

3.1 Issues of Control

If we begin with the issue of control, it appears, on the surface at least, that biobank regulation has empowered the individual donor very well. A plethora of international declarations, guidelines, directives, and some national laws have recently introduced an informed consent requirement in relation to tissue-based research. The Council of Europe's Convention for the Protection of Human Rights and Dignity of the Human Being with regard to the Application of Biology and Biomedicine from 1997 states, for example, that "When in the course of an intervention any part of a human body is removed, it may be stored and used for a purpose other than that for which it was removed, only if this is done in conformity with appropriate information and consent procedures" (Article 22). Further, in 2000, the World Medical Association, in the Helsinki Declaration (Article 1), updated its definition of medical research requiring informed consent to include "research on identifiable human material or identifiable data".

In reality, however, informed consent does little to provide donors with control over the samples they donate. The consent process typically constitutes a sort of fait accompli; a signature on the consent sheet is not the same as a positive influence over what the sample is used for. Furthermore, informed consent gives no measure of control once the sample has left the donor's body. The worries of donors concerning the abuse of genetic knowledge and the absence of social fairness are not likely to be meaningfully referred to in an information sheet. Informed consent is primarily of assistance to those who want to abstain from research participation. A set of confidentiality safeguards, however – namely the EU Directive 95/46/EC on Personal Data Protection and the US Privacy Rule from 2003 –, now provide donors with increased protection from unauthorized uses of data. Still, donors' sense of loss of control is very directly related to the necessity of institutional trustworthiness, not least for biobanks to address the issues of harmful uses and social fairness that are beyond the direct influence of the donors.

3.2 Issues of Harmful Uses of Medical Knowledge

With respect to the regulation of harmful uses of medical knowledge it becomes more complicated. 'Harmful uses' relate both to the regulation of research subject matter, i.e. what can be studied in the first place, and to secondary uses of the resulting knowledge (what is sometimes addressed with a so-called 'dual use clause' in ethics regulation). Is it possible to regulate the making and use of scientific knowledge? Most scholars would be inclined to answer both yes and no: "yes" because the social shaping of knowledge producing institutions does indeed influence the type of knowledge produced, as pointed out already by Max Weber (Weber 1947); and "no" because knowledge and technologies embody a transformative potential that can never be fully controlled. Once available, technologies may be used in unexpected ways by people other than those for whom it was developed (Oudshoorn and Pinch 2003). Take the example of Viagra, a pharmaceutical used to treat erectile dysfunction in men. It was developed while exploring the role of nitric oxides in the dilation of blood vessels in conjunction with heart disease, when its influence on erectile functionality was discovered; it was then marketed for middle-aged heterosexual men and subsequently appropriated by sexually hyperactive homosexual men as well as experimenting women as a booster of new sensual experiences (Mamo and Fishman 2001). Similarly, reproductive technologies developed to assist heterosexual couples with fertility problems have also been used by homosexual couples wanting to avoid certain sexual acts and family ties.

Regulation aimed at restricting research has at times been initiated by professional organizations, as in the moratorium on genetic recombination technologies in the early 1970s. At other times, national or supranational authorities have tried to ban particular types of research, as in the attempt to ban all human cloning by UN declaration (UN News Centre 2005) and national prohibitions on various types of research involving human gametes (Nordic Council of Ministers 2006). Most such bans and prohibitions are abandoned after a while. In an extensive study of regulatory bans on genetic research in the UK and the US, Susan Wright found that the impetus to national competition made most research bans short-lived (Wright 1994). An updated overview of the approximately 900 laws, regulations and guidelines governing human subjects research in 84 countries is published on an annual basis by the US Department of Health and Human Services, Office for Human Research Protections[1].

Concerning attempts to control the usage of some technologies there are examples of relatively long-lived prohibitions (e.g. abortion in some countries); or restrictions on legitimate user groups (as with reproductive technologies being available only for heterosexual couples in some countries) or attempts to protect vulnerable groups from exploitation (as with prohibitions on trade in organs). The authorization of medical treatment is another measure of control of the usage of research results, and thus a measure of protection of citizens. In the US, drug approval and inspection of tissue centres for therapeutic purposes fall under the jurisdiction of the Food and Drug Administration (FDA). Rules in both areas have been tightened considerably over the years. Drug approval regulation has been put in place to ensure testing on diverse user groups and to protect vulnerable populations from undue pressure. The tightening of the rules can be seen as a partial solution to the concerns donors express regarding equality in health. The full picture is however more complex than the emergence of stricter rules would suggest. In response to the new rules and corresponding increase in costs, the drug trial industry is currently being outsourced to less wealthy countries and populations (Petryna 2007). In Europe, attempts to stimulate innovation and research in the pharmaceutical sector have resulted in a harmonization of the national drug approval agencies and the establishment of the European Medicines Agency (EMEA). Competition between the US and the EU has been accused of changing the role of drug approval. Swift procedures have become a stated goal in order to attract the pharmaceutical companies who are now viewed as 'clients' rather than 'applicants' (Daemmrich 2004). The national EMEA agencies are typically dependent upon their clients for approximately 80% of their funding.

---

[1] http://www.hhs.gov/ohrp. Accessed 3 March 2008.

These trends reflect attempts to regulate the substance and subsequent uses of research. However, it has been argued that tools from administrative law such as authorization, prohibition and bans are of decreasing importance in the regulation of biotechnology. Instead, research of the type conducted on stored human tissue is regulated through other means, such as property law and funding regimes (Fleising and Smart 1993). Take the case of human embryonic stem cell (hESC) research: in the US, although this tissue-based research is largely unregulated in terms of direct prohibitions, it has received no federal funding since 2001. It has consequently become a matter for funding by the private sector, except for selected states that have initiated their own funding programmes. In the US, the patent practice in relation to hESC research has been very liberal and the patent application following the first isolation of hESC lines in 1998 (Thomson et al 1998) broadly covered the processes involved in the isolation as well. The breadth of these patents has been accused of impeding US stem cell research (Murray 2007). In contrast, hESC research in the UK is publicly funded, but more strictly regulated – though what is seen as strict in the UK is perceived as liberal in southern Europe. The European Patent Office (EPO) has held a more restrictive view on stem cell patenting than the US Patent and Trademark Office (USPTO) (Herder 2006, Plomer 2006). However, there is no agreement among the EPO countries about the patentability of, for example, embryonic stem cells (Comission of the Europiean Communities 2005). The UK authorities have argued that an embryonic stem cell line is not covered by the ban of patents on hESCs in the EU Biotechnology Directive (Plomer 2006). Political scientist Ingrid Schneider sees in the relative strictness on patentability in Europe a change in what patent law is expected to do (Schneider 2005). Whereas Fleising and Smart suggested that a general regulatory move from administrative law to property law was taking place, Schneider suggests that property law is now increasingly also seen as a research regulatory device (see also Gold and Caulfield 2002). The development of technology has been relegated to authorities that previously dealt only with private property interests – transfiguring the property system in the process. The combination of clear legal systems of authorization, generous funding, and a permissive approach to patentability has gradually secured the UK a position as the European hub of hESC research. Approaches to funding and patenting may therefore be just as important as legal authorizations and prohibitions for the development of a tissue-based technology.

3.3 Issues of Social Fairness

It is striking how the restructuring of funding and property regimes has, in most European countries, taken place during the same period in which the informed consent requirement was introduced to solve the problem of ensuring public trust in biobanking. These regulatory changes are characterized by an overall tendency to strengthen commercial incentives in the innovation system. Such changes are highly relevant to donors' concerns about social fairness, in particular whether research is dictated by creed rather than need, and whether everybody will get equal access to research results.

The inspiration for changes in the property regime has come from the US where the Bayh-Dole Act was passed in 1980. The Act transferred property rights from federal funding agencies to host research institutions, with the intention of encouraging universities and other research bodies to pursue a more active patenting and licensing strategy than government had previously succeeded in doing (Boettiger and Bennet 2006). In the subsequent period US patent applications and licenses have by far exceeded the numbers granted in Japan and the EU (Kortum and Lerner 1999). Japan and EU have gradually changed their innovation policies to compete with the US – in particular in the life science field. Intertwined with these changes the scope of patentability has also increased, again first in the US and then more reluctantly in the EU and Japan (Gold 1996). To facilitate the competitiveness of European universities and firms in the patent race with the US, the EU in 1998 passed Directive 98/44/EC on the legal protection of biotechnological inventions. At the same time, both the US and EU have gradually come to a different perspective on the patent application procedure. Like the EMEA and the FDA, the EPO and USPTO patent offices now refer to 'clients' rather than 'applicants' (Foray 2004). Court systems have also become increasingly friendly towards the patent holder: in the US only 62% of appeals by patent holders were successful in 1980, but by 2000 this figure had risen to 90% (ibid.). With international harmonisation initiatives, US patent policy has had repercussions for the life sciences worldwide (de Laet 2000). Gradually, elements from the Bayh-Dole Act have been implemented in the EU, encouraging commercialization of research. In Denmark, for example, ownership of property rights has been transferred from the inventor to the university, thus rewarding the funding agency for obtaining the patent. In Sweden, property rights still officially accrue to the inventor, but universities are rewarded for the establishment of successful spin-off companies in which the inventor shares the profit. These policy changes in regard to intellectual property rights influence the ability of universities and public hospitals to share material. As a result, the protection of commercial property interests has become a public duty. The new practice, in relation to tissue-based research, of signing contracts specifying commercial entitlements prior to any exchange of biomaterial is probably even more important than patent demands (Murray 2007). These so-called Material Transfer Agreements (MTAs) define the rights between the exchanging parties, but whereas a patent is publicly available the MTA contracts typically contain a confidentially clause.

Borrás argues that in the EU there has been a move from a science policy (focused on publicly funded basic research) to an innovation policy in which the role of public agencies is to support private initiative (Borrás 2003). Public research funds for biomedicine are therefore used differently, giving industry a greater influence on the research agenda. At a national level, many states have introduced measures into their funding policies to enhance the technology transfer from universities and university hospitals to pharmaceutical and biotechnology companies. A greater proportion of funds now demand co-financing from an industrial partner which, in effect, gives the partner a right of veto over the research agenda. The ambition of spending public funding more effectively by establishing quasi-markets for public research is particularly evident in the European welfare states (Le Grand 2003). The public sector is increasingly run by principles loosely known as New Public Management, and individual researchers must compete with each other rather than develop their research on the basis of continuous funding. US data indicate that ratios of public to private funding have remained more or less stable since the 1950s (Moses et al 2005). However, the amount may not be as significant as the conditions imposed on gaining access to the funding. When funding programmes require an industrial co-sponsor, the researcher relinquishes a certain degree of independence over the development of the research agenda. Finally, it should be remembered that other measures for creating competition among public researchers, while varying from country to country, typically include fiscal rewards dependent on rates of publication, ability to attract funding, patent applications, and other measures of productivity (Seglen 1997).

3.4 Summary

Control and protection
- Enhanced informed consent requirements
- Tightened rules for ethical approval
- Strengthened data protection
  - US Privacy Rule
  - EU Directive 95/46/EC on Personal Data Protection
- Prohibitions and bans, typically technology specific and temporary

Property law
- Bayh Dole Act and similar transfers of IPR to make public research institutions act more like companies
- Expanded property realm
  - US case law (Chakrabarty)
  - EU Directive 98/44/EC
  - International harmonization process

Research management
- Changed criteria for funding
  - EU encouraging public-private partnerships
- Quasi-markets created for public researchers
- Researchers and institutions rewarded for commercial initiatives

---

Textbox 1: Examples of regulatory changes in the innovation system in which biobanks operate

Textbox 1 summarizes the regulatory trends in the wider innovation system in which research biobanks operate. They are characterized by a paradoxical double move involving, on the one hand, stricter protection of research participants, including at least the right to abstain from participation, and, on the other hand, an intense reconfiguration of the incentive structure to reflect commercial objectives. The latter is rarely debated by ethicists but this reconfiguration is likely to influence the ability to meet donor expectations, particularly with respect to the issue of social fairness.

The following section will assess the impact of commercialization on medical research. I will argue that a commercialized incentive structure might indeed affect the issues of concern to donors.

4 What Does Commercialization Imply?

Most observers will agree that without the private sector few of the existing treatments for serious illnesses would exist. Not surprisingly, a recent report from the European Commission concludes that private sector involvement in medical research has a primarily positive impact on the number of products available (Zika et al 2007). Commercial interest in medical research is not only legitimate; it has proven itself to be productive and useful in several ways. Some critics of privatization – transfers from the public sector to the private sector – seem to have an alternative vision of medical research that is shielded from commercial interest altogether. I personally avoid the concept of privatization because the important feature for understanding the relevant consequences is what goes into the decision-making process rather than who employs the people making the decisions. Additionally, I think the vision of a totally non-commercial medical field is naïve. Rather, I use the term commercialization to indicate a transformation of the public research strategy that dissolves the distinction between public and private research by installing commercial incentives in all types of research. The question is what the current emphasis on commercial relevance in practically all decision-making regimes does to the research outlook. In assessing this, I will first try to sketch how commercialization influences research efforts in light of the stated goals surrounding the regulatory changes and then ask what additional effects commercialization is likely to have on the research agenda (what topics are being researched?) and the research results (is commercialization likely to generate pro-industry bias?).

4.1 Assessing the Impact in Light of the Stated Goals of Commercialization

If the reason for conflating public funding regimes with industry sponsorship was to promote technology transfer and, thus to enhance the number of actual products resulting from research, it would be interesting to know whether the goal was achieved. The methodological problem with reports like that of the European Commission mentioned above (Zika et al 2007), is the classical 'with/without' distinction: we are only aware of the products gained from the current sponsorship forms, not what would have been produced if a different funding and property regime had been in place. Still, there are some indications that industry sponsorship leads to a greater number of available products (i.e. a higher translation rate) than purely publicly funded research. The most significant breakthroughs, however, tend to use results of prior research based on a long preceding period of mainly public funding (Bouchard and Lemmens 2008). Mostly, the translation rate is low. In one study the authors identified all the articles in six major journals of basic science published between 1979 and 1983 in which it was stated that the technology had novel therapeutic or preventive promises (Contopoulos-Ioannidis et al 2003). They followed this up 20 years later. Despite choosing journals in which the most significant breakthroughs could be expected to be published, just 101 such articles were identified and of them only 27 ever led to a published trial. At the time of the study, just nine technologies had been in clinical use and four of these had been subsequently withdrawn from the market. Although it was found that industry sponsorship increased the likelihood of successful translation, given the limited number of successful cases, the statistical power of the results is low. What remains obvious is that it takes a lot of research, public or private, to produce a usable product.

It is more doubtful whether the amended property structures have succeeded in making universities and public hospitals operate more like profitable industrial partners. A handful of American elite universities have profited significantly on their patent portfolios, and they serve as an inspiration for European university politics. Furthermore, recent policy changes have stimulated increased patent activity; even if the EU and Japan still lag behind the US (Kortum and Lerner 1999). The relative American success might be related to the practical implications of different recruitment strategies in the European and US patent authorities: in the EPO countries the technical assessment of novelty of the application is conducted by people with technical training; in the US the investigation is conducted by people with juridical training, thereby implicitly leaving it to the courts to settle disagreements about infringement (Fernandes and Miska 2004: 255).

There may be reasons other than profit generation for filing patents (Hopkins et al 2006), but from the perspective of the strategic goals of commercialization it is worrisome that even in the US only one in 20 university patents is ever licensed, and only 5% of the licensed patents generate appreciable royalties (Fernandes and Miska 2004: 253). The European experience is even more depressing. Though filing more and more patents, European universities rarely make money on them. A UK study pointed to several reasons for this (Webster and Packer 1996a, Webster and Packer 1996b). Universities typically obtain only one patent in a given area, whereas a good bargaining position requires a structured and broad patent portfolio. Also, because universities are not present on the world market they do not detect infringements of their patents worldwide. When Denmark introduced Bayh-Dole like legislation some years after the UK, despite the disappointing UK experience, the results were more or less the same: the annual cost of claiming patents was calculated to be around 32 million DKK and the revenue about approximately 16 million DKK (Forsknings- og Innovationsstyrelsen 2006). It could be argued, of course, that it is not absolutely fair to judge the policies on their revenue if the overall purpose is to promote technology transfer and generate products. In this period of intensified commercialization, however, the number of scientific breakthroughs has gone down. Several commentators now complain that the innovation pipeline is drying up and they blame, in part, the increasingly short-sighted commercial focus embraced by basic research institutions (Bouchard and Lemmens 2008, Fernandes and Miska 2004).

4.2 The Influence of Commercialization on the Research Agenda

The distinction between short-term (often called 'applied') and long-term ('basic') research goals is also relevant to the influence of commercialization on the research agenda, i.e. the topics investigated. In the agricultural field it has been shown that the result of the modification of incentives for public research is that public research today mimics the private research agenda and no longer fulfils the earlier complementary function of investigating commercially doubtful research topics (Welsh and Glenna 2006). I know of no studies documenting the same tendency in the field of medicine. In one study, however, a group of researchers studied 3862 research careers to assess the impact of patent activity on scientific merits (Azoulay et al 2006). It was shown that commercial collaboration was associated with high scientific merits, but also with a different research agenda, involving systematic research in areas of commercial interest. Obviously, commercial partners must investigate what seems likely to generate profit. It is a matter of survival. Accordingly, when access to public funding becomes dependent on private partners willing to co-finance the projects, it is reasonable to expect that public researchers must also direct their interests to whatever constitutes a potential market.

Special funding initiatives have been put in place to counter this trend by prioritizing so-called orphan drugs – drugs for diseases too small to generate a reasonable market size. Also, special patent rules apply when a drug receives orphan drug status, and patent strategies have, therefore, been developed to market a drug at first related to an orphan disease and later to transfer it to be used for other purposes. The effect of the changes in funding and property regulation on the research agenda is therefore extremely difficult to assess. On the one hand, research must be directed to taking more account of market size; on the other hand, the bigger market can sometimes be attained by going first for something which also qualifies as orphan drug. The impact of increased patent activity has in itself been widely debated. Some talk about a tragedy of the anti-commons, meaning that too many patents impede research and leave certain promising research agendas unexplored (Heller and Eisenberg 1998, see also Andrews 2002, Cambon-Thomsen 2004, Goldman 2007) Others are far more optimistic about the ability of the patent regime to stimulate innovation (Kortum and Lerner 2007, Zikla et al 1999). The role of the MTAs is currently being explored, and clearly has an impact on the research agenda, in that projects without access to material need to change focus. In one Belgian study, 60% of the interviewed researchers said they had had to cancel projects because they could not negotiate access to the research materials they wanted (Rodriguez 2007, 360).

When discussing the influence of commercial incentives on the research agenda it is mostly taken for granted that if products can be sold, somebody must need them; hence, problems relating to changes in the research agenda are expected to concern either those products nobody wants to fund or equal access to new treatments. This analysis might be too crude, however, to capture the intricate interplay between product development and disease aetiology. It is commercially much more interesting to develop products that can be administered for longer periods of time (by wealthy people) to minimize potential risk of disease, for example, than to develop one-time cures or low-tech interventions which do not imply continuous use of any product. Therefore, the development of new "disease categories", such as hypertension, do not simply reflect discovery of a risk factor that can now be treated; such categories reflect a commercially potent interplay between discovery, assessment of public health needs, market size and new 'sick roles' for the not-yet-ill. Many new drugs are targeted at people who are in principle healthy but at risk, and marketing these products implies changing disease categories and the perception that people have of themselves as healthy. Additionally, when biobanks are used to identify means of prevention, industry-sponsored research is likely to try to identify modes of intervention (such as preventive drugs) that can be more easily capitalized upon, rather than addressing for example, elimination of risks from the environment or low-tech forms of preventive interventions.

Hence, commercial incentives might work contrary to the ambition of using stored tissue to address public health problems, both by impeding some research collaborations and by modifying the research agenda and our shared understanding as to what constitutes an important health problem.

4.3 The Influence of Commercialization on the Research Results

Finally, what do existing studies indicate in terms of influence of commercialization on research results? Is commercial affiliation leading to pro-industry bias? There is increasing evidence to support the claim that industrial sponsorship of a clinical drug trial is associated with product-friendly conclusions (Bhandari et al 2004, Stelfox et al 1998, Melander et al 2003, Jørgensen et al 2006). In a meta-study, 37 such studies where gathered (analysing a total of 1140 clinical trials), with a conclusion that was unequivocal: industry sponsorship is associated with pro-industry conclusions (Bekelman et al 2003). In another meta-analysis, 30 studies of the impact of industrial affiliation were analysed (Lexchin et al 2003). Again, the conclusion was clear: when sponsored by industry, research results are more likely to deliver pro-industry conclusions as well as to recommend more expensive treatment options. The statistical association between sponsorship and conclusions can be explained in various ways; for example, it has been argued that negative results are withheld from publication or simply delayed (Egilman 2005, Friedman and Richter 2005, Joly et al 2007). It is hoped that new systems for pre-trial registration will help to counter such suppression of publications (NIH 2007). It could also be that the association merely illustrates that industrial partners are skilled at choosing the products they want to put on trial. What it might imply for biobank-based research, in which no product is being tested, remains to be seen. But it should be obvious that there are good reasons to keep an eye on variations in research conclusions that might be associated with the source of funding.

Biobanks are used to identify determinants of health and illness, and there are clear industrial interests in downplaying or overstating certain associations. In a meta-analysis of 106 review articles on the health impact of passive smoking, for example, it was found that, according to 94% of the studies supported by the tobacco industry, no negative health impact of passive smoking was identified, while only 13% of the studies without industry support reached this conclusion (Barnes and Bero 1998). Furthermore, it has been claimed that epidemiological studies identifying risk factors with financial implications for selected industries tend to be de-bunked on methodological grounds by industry sponsored consultants (Pearce 2008). Precautionary measures are therefore postponed, sometimes for decades, until scientists can no longer 'disagree', given the number of studies in which the same results occur. Still, in most instances the industry has clear interests in reaching insight into correct associations. The pro-industry tendency in the accumulated research results remains a concern, however, not least because 87% of the authors of clinical guidelines also have industry affiliations (Caulfield 2007: 54), and the translation of evidence into clinical action requires reliance on the accuracy of all available studies, which might tempt reproduction of industry-friendly conclusions.

In short, a commercial reconfiguration of the medical research agenda and available results runs counter to the expectations donors have about medicine as a special field in which need rather than creed should determine action.

5 Conclusion

Trust cannot be taken for granted. Even where some degree of donor trust prevails it can only be preserved – and deserved – where it reflects trustworthiness. Taking trustworthiness to imply that the research infrastructure is designed to deliver what research participants expect it to, this chapter has compared what we know about donors' expectations to the overall regulatory tendencies governing biobank research.

The concerns of donors are largely related to their own lack of control over the tissue leaving their body; to harmful uses of medical knowledge and to issues of social fairness. Biobank legislation has introduced informed consent as a means by which the trust of the donating public should be preserved, and a number of legislatory changes have produced stronger data protection laws and certain bans and prohibitions that address concerns about harmful uses of knowledge. However, if it is correct to assert, as I have, that in order to preserve trust donor expectations must be met, more than data protection and a strengthened informed consent requirement are needed. As biobanks are embedded in wider innovation systems that shape their output, we must analyse biobanks in context and look further than the legislation that represents itself as governing biobanks. The wider innovation system is influenced by what I have called a tendency to commercialization, i.e. attempts to make public research institutions comply with what economists would call market forces. The current commercialization tendency is likely to impact on the ability of biobanks to meet the expectations of donors – in particular with respect to a research agenda that reflects public health needs and that is responsive to issues of social fairness. New policies are needed to address such issues. The recent interest in benefit sharing can be seen as one such attempt (HUGO Ethics Comittee 2002, see also Haddow et al 2007, Winickoff and Winickoff 2003). However, it is important to understand that the influence of commercialization runs deeper than what might be addressed by re-distributive politics. Health needs are co-produced with the knowledge we acquire about health and illness. A commercial research agenda affects the types of diseases we are able and willing to treat and the risk factors we detect. Of course, there is no research which is completely 'unbiased' in the sense 'not influenced by any interests'. Accordingly, we need multiple decision-making regimes in medical research in order to ask different types of questions concerning health and disease and to deliver different types of answers. Unfortunately, the current commercialization tendency represents a homogenization of decision-making criteria instead.

It is not a question of being altogether for or against commercial research. Public funding has typically been associated with high prioritization of military purposes – and private funding remains essential to the medical research field irrespective of our view of commercial incentives. The point is to use commercial incentives productively and to secure interests other than those that have a price (Caulfield et al 2006). Indeed, in light of the problems of empty pipelines and the lack of basic research, even commercial interests might need non-commercial modes of research sponsoring (Bouchard and Lemmens 2008). Therefore, commercial interests per se do not constitute a problem; it is the commercialization process, by which all decision-making regimes are directed toward the same type of (market-based) priorities.

The issue of trust is important, not only because biobanks will collapse if people stop trusting in them: it is important in order to show respect for the agency people exhibit through their donations. If donors do not read or recall their informed consent, then respecting their wishes requires more of researchers than legalistic scrutiny of the donor information sheet: the spirit in which donations were made must be respected. It might be an exception with purely commercial biobanks (Anderlik 2003, Lewis 2004), but the shared goals of the nation are typically appealed to when recruiting donors to participate in large-scale public databases, as pointed out by Helen Busby and Paul Martin (Busby and Martin 2006). A series of unspecified benefits are envisaged, but the ability to meet the generated expectations rarely receives the same attention (Petersen 2005). It is important, however, to begin to provide this attention and question the ability to deliver public goods. Changes in the research infrastructure are likely to influence the nature of the health benefits that follow from biobank research. Pointing to the contradiction between reasonable donor expectations of a research agenda built around public health needs and an increased emphasis on commercialization, Busby concludes that

> [...] a new agenda for medical research involving extensive public-private collaborations has emerged within a more traditional framework that fails to acknowledge conflicts, tensions and disjunctures between commercial interests and public sector research (Busby 2007: 184).

The exploration of these conflicts is an important task for ethicists.

The mismatch between social fairness and regulatory tendencies justifies a call for increased ethical scrutiny of the wider political and economical context of biobanking. Ethicists are often assigned a more restricted role by biobank organisers: they are expected to work out what should be included in the information sheet and when and by whom the consent form should be signed. If they wish to serve the interests of donors as well as biobanks, they must engage with a much broader set of issues. Ethicists must analyse the dynamics of the research infrastructure and evaluate its ability to accommodate what donors and the surrounding society can reasonably expect of a biobank project.

Important future work will require the input of ethicists, biobank researchers, economists, sociologists, funding agencies and legislators. It concerns the question of how to construct research infrastructures in a way that accommodates donors' concerns at the heart of the research endeavour. I contend that one element of this task will be to reintroduce some sort of distinction between commercial and non-commercial research. Attempts to counter the problem of a distorted research agenda must include the development of much clearer policies on licensing, royalties and other means of profit sharing for public research biobanks accessed by industry (Bouchard and Lemmens 2008). It will also include the development of clear priorities for the spending of such funds in ways that complement or perhaps counteract the commercial research agenda. Finally, special funding is needed to support research that challenges new categorizations of disease developed in commercialized research collaborations.

Biobanks are not just inconsequential collections of data and frozen blood samples. They are surrounded by many honest hopes and sincerely felt concerns. Cynical deliberation is known by the metaphor of cold-bloodedness; but in this instance the blood in the freezers should be treated with warmth and care. Trade is no problem, but trade in cold blood is. There is, therefore, a need to move beyond the consent issue in an attempt to make biobanks worthy of donors' trust.

Acknowledgments I would like to thank Mette Svendsen and Morten Andreasen for comments on an earlier draft of this chapter.

References

Anderlik MR (2003) Commercial Biobanks and Genetic Research: Ethical and Legal Issues. Am J Pharmacogenomics 3: 203-215

Andrews LB (2002) Genes and Patent Policy: Rethinking intellectual property rights. Nat Rev Genet 3: 803-808

Azoulay P, Ding W, Stuart T (2006) The impact of academic patenting on the rate, quality, and direction of (public) research. National Bureau of Economic Research, Cambridge

Barnes DE, Bero LA (1998) Why Review Articles on the Health Effects of Passive Smoking Reach Different Conclusions. JAMA 279: 1566-1570

Barr M (2006) 'I'm not Really Read up on Genetics': Biobanks and the Social Context of Informed Consent. BioSocieties 1: 251-262

Bekelman JE, Li Y, Gross GP (2003) Scope and Impact of Financial Conflicts of Interest in Biomedical Research: A Systematic Review. JAMA 289: 454-465

Bhandari M, Busse JW, Jackowski D et al (2004) Association between industry funding and statistically significant pro-industry findings in medical and surgical randomized trials. Can Med Assoc J 170: 477-480

Bister MD, Felt U, Strassing M, Wagner U (forthcoming) Refusing the information paradigm: Informed consent, medical research, and patient participation. Health

Boettiger S, Bennett AB (2006) Bayh-Dole: if we knew then what we know now. Nat Biotechnol 24: 320-323

Borrás S (2003) The Innovation Policy of the European Union: From Government to Governance. Edward Elgar, Cheltenham

Bouchard RA, Lemmens T (2008) Privatizing biomedical research – a 'third way'. Nat Biotechnol 26: 31-36

Busby H (2007) Biobanks, bioethics and concepts of donated blood in the UK. In: de Vries R, Turner L, Orfali K, Bosk CL (eds) The View From Here. Bioethics and the Social Sciences pp 179-193. Blackwell Publishing, Oxford

Busby H (2006) Consent, trust and ethics: reflections on the findings of an interview based study with people donating blood for genetic research for research within NHS. Clin Ethics 1: 211-215

Busby H (2004) Blood donation for genetic research: what can we learn from donors' narratives? In: Tutton R, Corrigan O (eds) Genetic Databases: Socio-ethical issues in the collection and use of DNA (p. 39-56). Routledge, London

Busby H, Martin P (2006) Biobanks, National Identity and Imagined Communities: The Case of UK Biobank. Sci Cult 15: 237-251

Cambon-Thomsen A (2004) The social and ethical issues of post-genomic human biobanks. Nat Rev Genet 5: 6-13

Caulfield T (2007) Profit and the Production of the Knowledge: The Impact of Industry on Representations of Research Results. Harv Health Policy Rev 8: 51-60

Caulfield T, Einsiedel E, Merz JF, Nicol D (2006) Trust, patents and public perceptions: the governance of controversial biotechnology research. Nat Bio-technol 24: 1352-1354

Clayton EW (2005) Informed Consent and Biobanks. J Law Med Ethics 33: 15-21

Commission of the European Communities (2005) Report From the Commission to the Council and the European Parliament. Development and Implications of Patent Law in the Field of Biotechnology and Genetic Engineering. COM (2005) 312. Brussels, The European Parliament

Contopoulos-Ioannidis D, Ntzani E, Ioannidis J (2003) Translation of higly promosing basic science research into clinical applications. Am J Med 114: 477-484

Cousins G, McGee H, Ring L et al (2005) Public Perceptions of Biomedical Research. A survey of the general population in Ireland. Health Research Board, Dublin

Cragg Ross Dawson (2000) Public Perceptions of the Collection of Human Bio-logical Samples. The Wellcome Trust and Medical Research Council, London

Daemmrich AA (2004) Pharmacopolitics. Drug Regulation in the United States and Germany. The University of North Carolina Press, Chapel Hill/London

Ducournau P (2007) The viewpoint of DNA donors on the consent procedure. New Genet Soc 26: 105-116

Earnscliffe Research & Communications (2000) Public Opinion Research Into Biotechnology Issues Third Wave. The Biotechnology Assistant Deputy Minister Coordinating Committee / Government of Canada, Ottawa

Egilman DS (2005) Suppression Bias at the Journal of Occupation and Environmental Medicine. Int J Occup Environ Health 11: 202-204

Fernandes M, Miska D (2004) Beyond Bayh-Dole and the Lambert Review: an Initial Product Development and Transactional Model for the Interface between Universities and Business. Biotechnol Genet Eng Rev 21: 249-276

Fleising U, Smart A (1993) The Development of Property Rights in Biotechnology. Cult Med Psychiatry 17: 43-57

Foray D (2004) The patent system and the dynamics of innovation in Europe. Sci Public Policy 31: 449-456

Forsknings- og Innovationsstyrelsen (2006) Kommercialisering af Forskningsresultater. Statistik 2005. Forsknings- og Innovationsstyrelsen, Copenhagen

Friedman L, Richter ED (2005) Conflicts of Interest and Scientific Integrity. Int J Occup Environ Health 11: 205-206

Gold ER (1996) Body parts. Property rights and the ownership of human biological materials. Georgetown University Press, Washington DC

Gold ER, Caulfield TA (2002) The Moral Tollbooth: A Method that Makes Use of the Patent System to Address Ethical Concerns in Biotechnology. Lancet 359: 2268-2270

Goldman,B. (2007) HER2 testing: The patent "genee" is out of the bottle. Can Med Assoc J 176: 1443-1444

Gudmundsdóttir ML, Nordal S (2007) Iceland. In M.Häyry, R.Chadwick, V.Árnason, & G.Árnason (Eds.), The Ethics and Governance of Human Genetic Databases (pp. 53-57). Cambridge: Cambridge University Press

Haddow G, Laurie G, Cunningham-Burley S et al (2007) Tackling community concerns about commercialisation and genetic research: A modest interdisciplinary proposal. Soc Sci Med 64: 272-282

Haimes E, Whong-Barr M (2004) Levels and styles of participation in genetic databases: a case study of the North Cumbria Community Genetics Project. In: Tutton R, Corrigan O (eds) Genetic Databases: Socio-ethical issues in the collection and use of DNA (p. 57-77). Routledge, London

Heller MA, Eisenberg RS (1998) Can patents deter innovation? The anticommons in biomedical research. Science 280: 698-701

Herder M (2006) Proliferating Patent Problems with Human Embryonic Stem Cell Research? J Bioeth Inq 3: 69-79

Hoeyer K (2003) "Science is Really Needed – That's All I Know". Informed Consent and the Non-Verbal Practices of Collecting Blood for Genetic Research in Sweden. New Genet Soc 22: 229-244

Hoeyer K (2004) Ambiguous gifts. Public anxiety, informed consent and commercial genetic biobank research. In: Tutton R, Corrigan O (eds) Genetic Databases: Socio-ethical issues in the collection and use of DNA (p 97-116). Routledge, London

Hoeyer K (2006) The power of ethics: a case study from Sweden on the social life of moral concerns in policy processes. Sociol Health Illn 28: 785-801

Hoeyer K, Olofsson BO, Mörndal T, Lynöe N (2004) Informed consent and biobanks: a population-based study of attitudes towards tissue donation for genetic research. Scand J Public Health 32: 224-229

Hopkins MM, Mahdi S, Thomas SM, Patel P (2006) The Patenting of Human DNA: Global Trends in Public and Private Sector Activity (the PATGEN Project). University of Sussex, SPRU

HUGO Ethics Committee. (2002) Statement on Human Genetic Databases. http://www.hugo-international.org/Statement_on_Human_Genomic_Databases.htm

Jack A, Womack C (2003) Why surgical patients do not donate tissue for commercial research: review of records. BMJ 327: 262

Joly Y, Wahnon F, Knoppers BM (2007) Impact of the Commercialization of Biotechnology Research on the Communication of Research Results: North American Perspective. Harv Health Policy Rev 8: 71-84

Jørgensen AW, Hilden J, Gøtzsche PC (2006) Cochrane reviews compared with industry supported meta-analyses and other meta-analyses of the same drugs: systematic review. BMJ 333: 782-785

Kettis-Lindblad Å, Ring L, Viberth E, Hansson MG (2007) Perceptions of potential donors in the Swedish public towards information and consent procedures in relation to use of human tissue samples in biobanks: A population-based study. Scand J Public Health 35: 148-156

Kettis-Lindblad Å, Ring L, Viberth E, Hansson MG (2006) Genetic research and donation of tissue samples to biobanks. What do potential sample donors in the Swedish general public think? Eur J Public Health 16: 433-440

Kortum S, Lerner J (1999) What is behind the recent surge in patenting? Res Policy 28: 1-22

de Laet M (2000) Patents, travel, space: ethnographic encounters with objects in transit. Environ Plan D 18: 149-168

Le Grand J (2003) Motivation, Agency, and Public Policy: Of Knights and Knaves, Pawns and Queens. Oxford University Press, New York

Levitt M, Weldon S (2005) A Well Placed Trust?: Public Perceptions of the Governance of DNA Databases. Crit Public Health 15, 311-321

Lewis G (2004) Tissue collection and the pharmaceutical industry: investigating corporate biobanks. In: Tutton R, Corrigan O (eds) Genetic Databases. Socio-ethical issues in the collection and use of DNA (p. 181-202). Routledge, London

Lexchin J, Bero LA, Djulbegovic B, Clark O (2003) Pharmaceutical industry sponsorship and research outcome and quality: systematic review. BMJ 326: 1167-1170

Malone T, Catalano PJ, O'Dwyer PJ, Giantonio B (2002) High Rate of Consent to Bank Biologic Samples for Future Research: The Eastern Cooperative Oncology Group Experience. J Natl Cancer Inst 94: 769-771

Mamo L, Fishman JR (2001) Potency in All the Right Places: Viagra as a Technology of the Gendered Body. Body Soc 7: 13-35

Melander H, Ahlqvist-Rastad J, Meijer G, Beermann B (2003) Evidence b(i)ased medicine – selective reporting from studies sponsored by pharmaceutical industry: review of studies in new drug applications. BMJ 326: 1171-1173

Merz JF (1997) Psychosocial Risks of Storing and Using Human Tissues in Research. Risk Health Saf Environ 8: 235-248

Moses H, Dorsey ER, Matheson DHM, Their SO (2005) Financial Anatomy of Biomedical Research. JAMA 294: 1333-1342

Murray F (2007) The Stem-Cell Market – Patents and the Pursuit of Scientific Progress. N Engl J Med 356: 2341-2343

National Institute of Health [NIH] (2007) Guidance on New Law (Public Law 110-85) Enacted to Expand the Scope of Clinical Trials.gov: Registration. Notice number NOT-OD-08-014

Nilstun T, Hermerén G (2006) Human tissue samples and ethics – attitudes of the general public in Sweden to biobank research. Med Health Care Philos 9: 81-86

Nordic Council of Ministers (2006) Assisted Reproduction in the Nordic Countries: A comparative study of policies and regulation. Nordic Committee on Bioethics, Copenhagen

Oudshoorn N, Pinch T (2003) Introduction: How Users and Non-Users Matter. In: Oudshoorn N, Pinch T (eds) How Users Matter – The Co-Construction of Users and Technologies pp 1-25. MIT Press, New Baskerville

Pearce N (2008) Corporate influences on epidemiology. Int J Epidemiol 37: 46-53

Pentz RD, Billot L, Wendler D (2006) Research on Stored Biological Samples: Views of African American and White American Cancer Patients. Am J Med Genet 140A: 733-739

Pentz RD, Young LN, Amos CJ et al (1999) Informed Consent for Tissue Research. JAMA 282: 1625

Petersen A (2005) Securing Our Genetic Health: Engendering Trust in UK Bio-bank. Sociol Health Illn 27: 271-292

Petryna A (2007) Clinical Trials Offshored: On Private Sector Science and Public Health. BioSocieties 2: 21-40

Plomer A (2006) Stem Cell Patents: European Patent Law and Ethics Report. University of Nottingham, Nottingham

Rodriguez V (2007) Merton and Ziman's modes of science: the case of biological and similar material transfer agreements. Sci Public Policy 34: 355-363

Schneider I (2005) "Taming the future with patents - frames and rhetoric in policy processes". Paper for ECPR (European Consortium for Political Research) conference, section 13: Theory and Praxis of Policy Analysis, September 8-10, 2005

Seglen PO (1997) Why the impact factor of journals should not be used for evaluating research. BMJ 314: 497

Shickle D (2006) The consent problem within DNA biobanks. Stud Hist Philos Biol Biomed Sci 37: 503-519

Skolbekken J-A, Ursin LØ, Solberg B, Christensen E, Ytterhus B (2005) Not Worth the Paper it's Written on? Informed Consent and Biobank Research in a Norwegian Context. Crit Public Health 15: 335-347

Stegmayr B, Asplund K (2002) Informed consent for genetic research on blood stored for more than a decade: a population based study. BMJ 325: 634-635

Stelfox HT, Chua G, O'Rourke K, Detsky AS (1998) Conflict of Interest in the Debate over Calcium-channel Antagonists. N Engl J Med 338: 101-106

Sutrop M (2007) Trust. In: Häyry M, Chadwick R, Árnason V, Árnason G (eds) The Ethics and Governance of Human Genetic Databases (p. 190-198). Cambridge University Press, Cambridge

Svendsen MN (2007) Mellem reproduktiv og regenerativ medicin. Donation som handlerum i fertilitetsklinikken. In: Koch L, Hoeyer K (eds) Håbets teknologi. Samfundsvidenskabelige perspektiver på stamcelleforskning i Danmark (p. 176-200). Munksgaard, Copenhagen

Thomson JA, Itskovitz-Eldor J, Shapiro SS, et al (1998) Embryonic Stem Cell Lines Derived from Human Blastocysts. Science 282: 1145-1147

UN News Centre. (2005) General Assembly Approves Decleration Banning All Form of Cloning. UN News Center, New York

Weber M (1947) Science as a Vocation. In: Gerth HH, Mills CW (eds) From Max Weber: Essays in Sociology (p. 129-156). Oxford University Press, New York

Webster A, Packer K (1996a) Intellectual Property and the Wider Innovation System. In: Webster A, Packer K (eds) Innovation and the Intellectual Property System pp 1-19. Kluwer Law International, London

Webster A, Packer K (1996b) Patens and Technology Transfer in Public Sector Research: The Tension Between Policy and Practice. In Kirkland J (ed) Barriers to International Technology Transfer (p. 43-64). Kluwer Academic Publishers, London

Weldon S (2007) United Kingdom. In: Häyry M, Chadwick R, Árnason V, Árnason G (eds) The Ethics and Governance of Human Genetic Databases (p 66-72). Cambridge University Press, Cambridge

Welsh R, Glenna L (2006) Considering the Role of the University in Conducting Research on Agri-biotechnologies. Soc Stud Sci 36: 929-942

Wendler D (2006) One time general consent for research on biological samples. BMJ 332: 544-547

Wendler D, Emanuel E (2002) The Debate over Research on Stored Biological Samples: What Do Sources Think? Arch Intern Med 162: 1457-1462

Winickoff DE, Winickoff RN (2003) The Charitable Trust as a Model for Genomic Biobanks. N Engl J Med 349: 1180-1184

Womack C, Jack A (2003) Family attitudes to research using samples taken at coroner's post-mortem examinations: review of records. BMJ 327: 781-782

Womack C, Pope J, Jack A, Semple C (2006) Cadaveric Tissue Retrieval Service for Research: One-year Review and Options for the Future. Cell and Tissue Banking 7: 211-214

Wright S (1994) Molecular Politics. Developing American and British Regulatory Policy for Genetic Engineering 1972-1982. University of Chicago Press, Chicago

Zika E, Papatryfon I, Wolf O, Gómez-Barbero M, Stein AJ, Bock A-K (2007) Consequences, Opportunities and Challenges of Modern Biotechnology for Europe. Spain, European Commision, Institute of Prospective Technological Studies

Biobanking

Trust as Basis for Responsibility

Cornelia Richter

Abstract In this paper, it is argued that far from simply being a problem within scientific research, biobanking is indeed an issue of societal debate, including various aspects of risk and safety, responsibility and trust. Being a highly complex issue with international and (inter-)cultural impact, biobanking turns out to be a challenging field of tensions from an ethical and even philosophical point of view. Especially those features which at first sight seem to be related to technical safety and thus technical improvement only, turn out to be inevitably bound to historical and theoretical notions of risk, safety and security, which have to be connected to the notion of responsibility. Following this line of argument, this paper will turn to the notion of trust (in biobanking) – a concept which has ever since been closely connected to instances of risk, safety and responsibility: Firstly, because trust implies the idea that a successful outcome is neither predictable nor can it be demanded. Instead it is aimed at basic openness and as such strictly distinct from any legal contract. But this is, secondly, exactly the reason why trust depends on at least any sort of stability, gained from intersubjective and social consensus but not restricted to socially traditional forms of life. Trust is built on recognition; it may evolve when people think and act similarly or loyally support each other's otherness. Both aspects, however, point to one basic human need, namely the need for security in its broadest sense, implying different features such as stability, accountability, certainty and thus the ability to act confidently. "Trust" therefore has to be understood as something which demands knowledge and consent while still taking experiences of uneasiness and fear seriously and thus respecting the basic need for security. It is highlighted that – in order to gain trust as precondition for success – biobanking indeed has to prove itself trustworthy.

1 Introduction

Already in 2004 Anne Cambon-Thomsen noted a peculiar shift in public attendance towards biobanks: "From being clinical or academic research tools that were largely ignored by the general public, they [i.e. biobanks, CR] have become a subject of societal debate. They have acquired the status of national resources" (Cambon-Thomsen 2004, 866). Only one year later J. Harris stated that science itself was "under attack", more than once being labelled as "Frankenstein science" (Harris 2005, 245). There is little doubt that Cambon-Thomsen and Harris are right but the questions would then have to be: Which features characterize this societal debate? And why is it lead in such an agitated way?

If we have, as Cambon-Thomsen does, the term "biobank" refer "to organized collections of biological samples and the data associated with them", which "come in many different forms, according to the type of samples that are stored and the domain in which they are collected" (Cambon-Thomsen 2004, 866), the multiple aspects of the problem become obvious: Being a project of medical research, scientists invest their knowledge, capacity, passion for many years in extremely expensive projects, which are then opened for widespread research by others. First of all, this implies all sorts of technical and scientific issues: Things simply have to work, technical processes in laboratories, storage facilities, databases and evaluations have to be carried out and maintained in an absolutely reliable manner, there has to be enough material supplied and qualified staff to handle it properly. Precautions have to be taken to make sure the data will be handled according to high standards, efficiently and responsible – otherwise all work would have been in vain. Yet, what at first sight seems to be a purely *scientific*, even *medical and technical* debate is in fact a *psychological, educational, social and institutional* one as well: Biobanks contain data of thousands of individual patients, subject to their consent. Individual patients have to answer to the questions with full responsibility, with help of their doctors at home or in hospitals. Scientists depend therefore on both the cooperating doctors as well as the patients' answers. They have to trust their intellectual and emotional abilities to understand the questions, their honesty in providing embarrassing details, especially concerning those aspects of daily life which are well-known to be unhealthy and unwise (alcohol, smoking, fast food etc.). The patients, however, depend on the reliability of the research project concerning the guarantee of privacy of their data. Either they are rendered anonymous or they are treated with even more care. There have to be clear rules about who will get access to these data, especially those referring to future prognostics. This is even more precarious when biobanks are managed and/or financed by private institutions which cannot be held responsible in the same way as national public institutions can, which finally brings us to *economic and political aspects*: What happens, if the institutions in charge simply collapse, if they file for bankruptcy or if political stability is no longer given? In any case, those who fund biobanks, nations or private investors, have to trust their management in both, their ability to cover financial expenses and take preliminary steps to keep them covered in the future as well as keeping them flexible enough to react to and at the same time withstand political changes or even crises. Furthermore, the large-scale collections of sensible medical and other data are gathered in instances of severe illness. Scientists are convinced that the collections will enable us to research in incomparably more detail than it has been possible so far. Especially diseases like cancer, diabetes etc. will be better understood they hope. Since we know that most diseases bear relevant hereditary aspects as well as they are reactions to certain ways of life, mainly unhealthy ones, long-term studies may enable us to better predict and prevent such diseases. However, only if the amount of data is large enough to also mirror geographical, climatic and regional, gendered, age-related and cultural habits such as nutrition, lifestyle etc. will they provide us with a full picture of elements facilitating or hindering the development of life – which finally makes the whole issue a question of *international and (inter-)cultural impact* as well.

Hence, biobanking is without doubt a highly complex issue, having each of its features imply several possible and serious fractures of the system. Once projects have reached a certain level of (global) complexity, even those features which at first sight seem to be related to technical safety and thus technical improvement only, turn out to be inevitably bound to and dependent on the whole setting. So, in this paper, I will argue that far from simply being a problem within scientific research, *biobanking is indeed an issue of societal debate, including various aspects of risk and safety, responsibility and trust*. There is indeed a high risk to biobanking; it has to be taken care of with great responsibility, striving for as much safety as possible – a declaration that seems to be nothing but self-evident. Yet, it is already from the history of these concepts that we can learn about their critical, even precarious mutuality. So in order to better understand the tasks at hand, I will therefore *firstly* present some historical and philosophical aspects of the notions of risk, safety and security, which will *secondly* be connected to the notion of responsibility. Although the notions of risk, safety and responsibility have always been related in some way or another, each of them has a specific history of its own. The development of concepts and terms is without doubt always distinct from the matter they refer to – so there naturally have been considerations of risk, need of safety or theories on responsible acts long before. But the emergence of a certain topic in a precise term is something different as it indicates a significant change of usage. In our case all three notions stem from rather different contexts but became legal and political concepts around the 14th and 15th century. All of them had to be critically revised first in the process of Enlightenment, then once more in the big European cultural crisis around 1850, and in the 1970ies they finally became irreversibly connected in the discussions on oil crisis, nuclear technologies, economics and other global problems. In the *third* part I will turn to the notion of trust as the discussion appears to concentrate on exactly this question whether we trust in biobanking or not. Considering the line of thought presented above, however, this does not at all come as a surprise. The concept of trust has ever since been closely connected to instances of risk, safety and responsibility. Firstly, because trust implies the idea that a successful outcome is neither predictable nor can it be demanded. Instead it is aimed at basic openness and as such strictly distinct from any legal contract. But this is, secondly, exactly the reason why trust depends on at least any sort of stability, gained from intersubjective and social consensus but not restricted to socially traditional forms of life. Trust is built on recognition; it may evolve when people think and act similarly or loyally support each other's otherness. Both aspects, however, point to one basic human need, namely the need for security in its broadest sense, implying different features such as stability, accountability, certainty and thus the ability to act confidently. So in the end we are back to the complex, intertwined and interdependent relation between trust and risk, safety, security and responsibility.

2 Risk, Safety and Security

The *concept of risk* first arose, as Otthein Rammstedt showed, among Italian merchants in the 14th century as "risco" or "rischio" (cf. Rammstedt 1992, 1045) and was primarily used in maritime insurance, referring to the danger of possible damage to cargo ships, storage or ongoing trades. So 'risk' was used to name the fact *that* an unwanted event might occur upon which preventative steps in form of financial insurance could be taken. Here, economics were based on formal rationality, mirrored in the calculable monetary system (cf. Rammstedt 1992, 1046). Up to the 20th century 'risk' remained a mainly commercial term but the idea behind it was soon strongly influenced by all versions of probability theory, especially by the efforts of Blaise Pascal (1623-1662), Pierre de Fermat (ca. 1607-1665) and above all Pierre-Simon Laplace (1749-1827). Whereas until Laplace it seemed to be impossible to bring something as contingent as reality down to mathematics he, in his "Théorie Analythique des Probabilités" (1812), showed that the analysis of random phenomena could provide relevant statistic patterns for further examination to draw upon. Both lines of thought were finally taken up by economic theory, e.g. by Adam Smith (1723-1790) and F.H. Knight (1885-1972). In his famous study "Risk, Uncertainty and Profit" (1921) Knight developed a theory of risk by analysing different modes of probability and by differentiating between "risk" and "uncertainty" which he described to differ in terms of measurability: Risk he used for "measurable uncertainty", "uncertainty" for "unmeasurable uncertainty" (Knight 1964, 233) which allowed him to eliminate the ambiguity implied in the way both terms were ordinarily used, namely be differentiating them in terms of favourableness. This would allow to understand insecurity relating to future developments and data not as something one would want to overcome as soon as possible in order to find solid ground again. Instead Knight declared risk an essential part of economy as there was no profit to gain without risk – economy being the very complex and unstable thing it is.

It is interesting to see in which way the notion of risk changed with time from the rather unspecific notion of "an unwanted event which may or may not occur" (cf. Hansson 2007) to the more exact concept of risk as "the cause" of an unwanted event which may or may not occur or of risk as "the probability" of such an event to occur (ibid.). The core of this change is the increasing accent on the *calculability and accountability* of the event one fears to happen. From our perspective it is intriguing to see how, especially in Knight's words, this decisive approach to reality being uncertain and shattered obviously mirrors a specific spirit of that time. For it was also in the 1920s that philosopher Martin Heidegger (1889-1976) developed – hereby following the works of Søren Kierkegaard (1813-1855) – the idea that being itself was venturous and that therefore nothing except anxiety could give access to one's real self (cf. Rammstedt 1992, 1047f.). In his excellent article in "The Stanford Encyclopedia of Philosophy" Sven Ove Hansson extended this line of economic theories up to Markowitz/Tobin in the 1950s, Rothschild/Stiglitz in the 1960s and Tversky/Kahnemann in the 1980s, but for our concerns another point of view has to be taken.

Since what in economic concerns may have been challenging and promising fortune turned out to be life-threatening in other contexts – and it is here that the notion of risk becomes closely tied to that of safety and security: In the 1970s it became evident that nuclear technology would pose new questions beyond Hiroshima and Nagasaki. For now the threat of nuclear contamination forced politics to decide on technical-economical risks in a rather ambivalent way: How much safety would be economically reasonable? How much insecurity could society be expected to bear? How should the risks of nuclear power be evaluated and how should they be taken care of – dangers that would neither respect national borders nor individual evaluation (cf. Rammstedt 1992, 1048)? In "Reactor Safety Study of 1975 Rasmussen et al provided the standard definition of technological 'risk' being valid until today: The expectation value of a possible negative event is the product of its probability and some measure of its severity (according to Rammstedt 1992, 1048). Whereas, in 1969, C. Starr could still plead for weighing up reasonable evaluation of risk versus individually calculable bargain, evaluation of subjective and objective risks became more and more complex: "[S]ubjective appraisals of risk", as Hansson puts it, "depend to a large extent on factors that are not covered in traditional measures of objective risk" (cited by Hansson 2007). Thus evaluation of risk is scarcely based on individual decisions but carried out in the expectation of technical safety (cf. Rammstedt 1992, 1048).

Although the history of the *notion of safety* in its results bears striking semblances to that of risk, it evolved in a completely different way. Whereas risk had always been more or less confined to economic processes, the notion of safety resp. security covered a much larger range of meaning. Different to 'risk', 'security' is an old term, going back to the epicurean idea of "ataraxia" and known since the first century as "securitas", i.e. a state of mind free of pain and suffering, enabling fortunate life (cf. Makropoulos 1995, 745f.). In medieval times 'security' first became a political and legal notion covering the idea of peaceful life within defined borders on the one hand and being part of all sorts of legal contracts securing one's life and property on the other. In the 16th century security became an essential concept in political theory, especially in Machiavelli (1469-1527), employed to describe the balance between international political powers and inner governmental control (cf. Makropoulos, 747). In the development towards Enlightenment, however, it was exactly the latter aspect that turned out to be highly ambivalent: The idea of a bourgeois society of free and responsible individuals was no longer consistent with the enactment of absolute power and control by the sovereign. Yet bourgeois society was nothing to easily establish; in fact it demanded a long process of sorting out legal and moral systems which would keep humans from simply taking advantage of their freedom. 'Insecurity', as Makropoulos put it, therefore had to be defined anew as endangerment of man through man (cf. Makropoulos 1995, 747). Hence, security could be understood, as Thomas Hobbes (1588-1679) said, as "not only their Consent, but also the Subjection of their wills in such things as were necessary to Peace and Defence; and in that "Union and Subjection", the nature of a *City* consisted; [...] for security is the end wherefore men submit themselves to others" (Hobbes 1983, 93; VI/3). Safety [NB!], however, was "not the sole preservation of life in what condition soever, but in order to its happines" [sic!] (Hobbes 1983, 158; XIII/4). In this line security finally became the leading idea for saving the individual's ability to self-development by securing the rights of all individuals likewise. Hence, according to Immanuel Kant (1724-1804) whose philosophy called for the autonomous, responsible individual, it was governmental duty to provide a constitutional state instead of acting as police state – an idea prevalent until the 19th century which the Austrian writer Stefan Zweig (1881-1942) named the golden age of security (cf. Makropoulos 1995, 748).

It is in many respects that the 19th century was called an extremely successful era: The rapid development of sciences and economics made industrialization and capitalism the leading ideas, bringing along mobility, urbanization and thus new forms of social life. One of the highlights of the time was the 1st world exposition in London in 1851, starring the engineer as prototype of the responsible individual, with clear intent and acting upon it. Yet, at the same time the other side of the coin also became visible, the golden age being replaced by the impression of a deep European cultural crisis, starting around 1850 and finally culminating in World War I. Due to the rise of capitalism, bringing along mass production, the proletarian masses, poverty, social friction etc., the idea of autonomous subjectivity became shattered. Instead there seemed to be an increasing gap between individual action and responsibility as the former causal relation was no longer given. Mass production demanded division of labour, depending on conditions and lead by intentions a-present or even unknown to those actually carrying it out. But who, then, was to blame in case of danger? For his own life Stefan Zweig thus remarked, one had learned to live without standing on firm ground, living without security (cf. Zweig 1982, 14 and 18) – hereby coming close to Knight and Heidegger who, as we have seen above, took risk and venturous being for granted.

And indeed, from now on, the notions of *risk and security* met in a peculiar way as the concept of security became more and more important for the prevention of social and technical dangers in the face of global threats, accentuating the concept of safety. While in German both aspects are covered in the word "Sicherheit", the English language allows distinguishing between the social and political notion of security on the one side and the more technical notion of safety on the other. Hence, endangerment was no longer consigned to other individuals or groups of individuals but gradually turned into abstract claims for social security or technical safety, measured by statistical factors of dysfunctionality and the awareness of the dangerous dynamics of risk (cf. Makropoulos 1995, 749). Most prominent among the manifold political programs was the 1948 United Nations claim for the right to social security. But, as we have seen above, in post-war Europe it became clear that all national efforts for inner security, which now had to include questions of environmental and social politics as well, would not achieve much. Instead the decision for oil or nuclear technology demanded completely new perspectives on security and safety as they were no longer to be solved by single nations let alone by the individual. Security and safety were divided into pieces because they became diverse and complex, relating to and depending on objective aspects of quite a different kind. Consequently the reaction to this process was complaining about the loss of old norms and values, not only by individuals but also in political programs: "Sicherheit", security as well as safety, thus became a key concept – indicating a deep societal problem (cf. Conze 2005).

3 The Concept of Responsibility

It won't come as a surprise to see that the notion of *responsibility* shares the historical development outlined so far in many ways. Just like risk and security/safety, responsibility seems to be a concept almost self-evident. Yet, surprisingly little research has been done on the history of this basic idea and opinions differ widely as to what is captured by the term. Kurt Bayertz (Bayertz 1995) has portrayed some of the big historical changes concerning this idea: Similar to 'risk' the notion of responsibility is (at least in the German speaking countries) first to be found around the 15th century, mainly in legal contexts though not in a commercial sense. Earlier versions of the idea, e.g. in Aristotle's' "Nicomachian Ethic" also thoroughly reflected on responsible acts but in a completely different way, representing the pre-critical stages of the notion we have already seen in the concept of security. Aristotle would understand responsible acts as causal relations between acting individuals and the objects acted on, producing certain effects. His aim was to reflect upon the conditions for the individual to act, e.g. whether s/he acted voluntarily and in full consciousness of all circumstances or not. If not, the action could not be counted as responsible act (cf. Aristotle 1985, 48). The significant change in the notion of responsibility happened in connection with the modern usage of "persona" as a subject being accountable for, but it was not before 1850 that 'responsibility' became a key concept, i.e. in the time of cultural crisis in Western Europe. The classic distinction between causality, intentionality and individuality of actions which until then had been valid became problematic. According to this distinction the individual sought to justify his/her deeds before his/her own conscience. The latter was regarded autonomous, following the normativity which rationality provided – just like Kant had suggested. In certain respects this idea might have worked well in Enlightenment, but it could definitely not be upheld in times of mass production as we can see for example in Burke's famous analysis of bursting boilers in the USA. He showed that the damage could not be blamed on the single engineer but on the whole technical process and had thus become a public problem (cf. Bayertz 1995, 26f.). So strategies of prevention became central and things became increasingly worse with global developments, particularly with the danger of nuclear technology.

Consequently there has been a crucial turn in the concept of responsibility: On the one hand the concept became increasingly formal as it was used for both, for in deed "responsible" acts of the individual as well as for any competence, responsibility or powers ("Zuständigkeit") of anybody for any task at hand (cf. Bayertz 1995, 32). Hence, in the long run 'responsibility' became a functional and even de-moralized term which is nothing but the exact opposite of its former classical version. Moreover, Bayertz claims, the concept would thus lose its clear direction to a certain problem and would be used for a great variety of social problems instead, among which the question of the accountability of negative consequences of events or acts was but one question among others (cf. Bayertz 1995, 42). On the other hand this development has lead to strong ethical demands for responsibility as we can for example see in the works of Hans Jonas. In his famous book "The imperative of responsibility. In search for an ethics for the technological age" (1984, first published in German in 1979 under the title "Das Prinzip Verantwortung") Jonas demanded a complete revision of ethics as traditional models and moral philosophies had become void in relation to modern technologies and global developments. Especially the Christian idea of charity had had its day as our "neighbour" today would live in global distance. So in order to escape pure existential anxiety and harsh critique of civilization new ways of orientation had to be found and new "categorical imperatives" had to be established. But in fact Jonas hereby simply reinstates individual rationality as the one and only principle of constructing responsibility, almost exempt of time. Hence he seems to be almost ignorant of the historical development of morality and ethics (cf. the critique of Bayertz 1995, 32ff. and 48-64). According to Bayertz, and I tend to agree with him, such new approaches to the concept of responsibility do not solve but express the problems of coordination and control which result from complex organizations of division of labour in modern societies (cf. Bayertz 1995, 34).

Given this short review of the conceptual history of risk, security, safety and responsibility, simple demands for responsibility obviously fail – at least when it comes to highly complex issues such as biobanks. Even though all efforts could be made to carry out all technical processes involved in laboratories, storage facilities, databases and evaluations as safely as possible, even though one would seek to provide enough material to supply, and even though members of staff could be trained even harder to handle things properly, few of the problems involved here could in fact be solved by technical or administrative progress alone. What used to be accounted for in terms of objective risk, provided security and individual responsibility, has merged into a conglomeration of medical, technical, scientific, economical, social, moral, international and intercultural aspects, having each of its features imply several possible and serious fractures to the system so that questions of risk and security will inevitably remain part of the game. Hence, responsibility has to draw on other resources in order to still be taken on – and it is here that the notion of trust comes into play; not "even though" but "exactly because" accountability of objective risks or individual responsibility for security as well as for safety seem to be widely lost. New and different forms of social interaction, due to scientific progress but also due to big changes in gender-, age- and generational politics, education and job-perspectives, mobility and so forth, have produced a phenomenon I would like to call *enforced individuality*. As an example one might mention new technological processes in prenatal medicine, especially in early diagnostic and therapeutic approaches such as screening models, 3D ultrasound or new neuroimaging techniques. They allow us far better diagnostics but they also crucially enforce doctors and parents to decide which forms of life they consider worth living. To name but another example, in case of illness we can choose between several specialists, hospitals and therapies on which most of us are able to gather qualified information on the internet – but in the end there will be nothing left for us but to trust the doctors. Strategies of coping and seeking responsibility demand inner as well as external freedom to leave behind what still seems to be solid ground (or is at least known as such). So in order to understand why the debate on biobanks and issues of similar complexity is lead in such an agitated way we have to inquire not only about new perspectives and revolutionary scientific developments but also sources and backgrounds of experiences of uneasiness, feelings of uncertainty, anxiety and even threat.

4 Trust – in Need of Security

And indeed, as I have stated above, the concept of trust has ever since become a key notion as issues of risk, security, safety and responsibility have become crucial. A closer look at present theories on trust reveals that many of them deal with problems in the context of politics, technical development and economics. In his excellent study "Trust. Reason, Routine, Reflexivity" from 2006 Guido Möllering has presented an overview on the debate under the core question: "What makes trust such a powerful concept?" (Möllering 2006, 6). In his answer he also points to the "inherent ambivalence of trust as the main feature that makes the concept so interesting and unusual. Trust has both highly uncomfortable and highly positive connotations." (Möllering 2006, 6) On the one hand, trust seems to be lost in accordance with loss of traditional structures in sociality and public life; on the other hand, this is regarded as exactly the reason for an increasing need for trust (cf. Möllering 2006, 2). But in any way, it is a highly complex phenomenon: "[W]ithout actors, expectations, vulnerability, uncertainty, agency and social embeddedness, the problem of trust does not arise and, if this were the case, the conceptualization of trust would be pretty meaningless or superfluous." (Möllering 2006, 9). In this section I will therefore first present some basic historical lines of argument (4.1), setting the stage for a significant change during Enlightenment (4.2), and finally outline two major lines of thought concerning 'trust', namely social and political interaction on the one hand and the concept of nature on the other, which have influenced the debate until today (4.3).

4.1 Historical Background

It seems that trust has all along been experienced in a rather diversified way, mostly kept in the tension between loyalty and faithfulness, reliability, affective emotion, hope and belief (cf. to the following aspects Gloyna 2001). Already in antiquity the notion of trust was used to describe basic preconditions of intersubjective communication and action, especially in legal and political contexts. Aristotle for example held that *polis* could only function based on trust among its members whereas *tyrannis* could only be maintained on the basis of general mistrust and anonymity. As soon as people would engage and interact in close and honest relationship, their loyalty would prevent despotic systems (Aristotle, 1990, V/11, 1313 a 34 – 1314 a 29). Such noble characteristics, however, have to meet with noble characters which made antique authors claim that trust was a matter of individual self-respect and reputation, which Cicero described as *fiducia*, fundamental for both becoming a brave soldier as well as a reliable business partner. In the latter sense *fiducia* was even part of Roman civil law, namely in the so-called *Formula Beatica* "fidi fiduciae causa" (cf. Gloyna 2001, 986).

It is interesting to find a similar line in the Old Testament, even though there is no single word for "trust" but several similar meanings instead (cf. Jepsen 1973). In most cases the texts speak of faithfulness, reliability and permanence of single persons who as such stand out and help to restore or keep the security and peace of the community. Usually, however, utmost caution is called for as it is difficult to decide whether people are reliable or not. Hence, several texts of the Old Testament express deep understanding for cautious consideration or careful distrust in claims for trust, which is even more understandable considering the oriental nomadic societies of that time. Even a leading figure such as Moses has to fear that people won't trust his words – a fear which God himself seems to share and therefore "supplies" Moses with signs and miracles (e.g. Ex. 4). The other way round, however, in relation to God, trust is used to characterize God as the one and only who will keep his promise for ever, who will keep his Word. In fact, he *is* "trust" in the very sense of the word as "nothing is as sure, permanent, or reliable as God" (Blackman 1982, 222). What is captured in the Hebrew word "amn", one of the prominent words for "trust", is until today implied when prayers are ended with "Amen", confirming the reliability and truth of what has been said.

In the New Testament this latter notion is still valid, but here its objective is clearly Jesus Christ. According to the Synoptic Gospels, Jesus himself placed his trust in God whom he addressed as his Father, throughout his life being deeply convinced of God's loving presence and wisdom, and whom he obviously felt bound to in a unique relationship – unique to such an extent that it was later expressed in the Christian doctrine of his divinity and unique sonship (cf. Blackman 1982, 229). So in early Christianity "faith meant, in the first place, acceptance of the gospel message", namely "that God's redemptive actions culminated in Jesus of Nazareth, whose divinely controlled ministry terminated in martyrdom, but who was authenticated as Messiah and Lord by his resurrection from the dead (Acts 2, 36-38)." (Blackman, 1982, 230) The interesting move we find here is that the general subjective notion of trust changed towards the concentration on objective contents, namely the Christian gospel. Especially the writings of Paul and the Gospel of John have contributed to this development: Paul as he considers faith as precondition, as starting point for our relation to God, but not as its goal, growing out of our experience with God. Hence, in and through faith we completely change our approach to the world and to others, to relationships, actions and values. John even sharpens the notion of faith to a notion of knowledge, namely the knowledge of God which is revealed by Christ, the Word and Son of God. Different to Paul, John understands faith as being raised by seeing and witnessing signs of God's power in what Jesus said or did. Both interpretations, however, seem to be in full accordance with how Jesus himself had lived: For throughout his entire life, it seems to us, Jesus was convinced of God's loving presence and wisdom. Faith in God would help to survive; faith in God would make people see; faith, however, which could be learned from Jesus, in following him and his words. But then we certainly have to ask: Isn't it true that, in the end, even Jesus himself died in despair, crying? And if this happened to him, who was called the Son of God, how should we ever share the optimistic claim of God's eternal love? How should we be able to trust in God facing severe illness or the loss of a child? But interesting enough, Paul's considerations of faith do imply this dimension as well: Whatever we do or say or think is useless unless it is done, said or thought in full recognition of God's love. Neither wishes nor prayers – like in the last hours of Jesus' life – or other efforts are of any avail. Instead, and this is exactly what we can learn from Jesus, "[m]an must simply confess his helplessness and make himself open to the divine grace. This fundamental humility and willingness to depend on God, abandoning self-sufficiency and the effort to make oneself worthy, is faith in the Pauline sense." (Blackman 1982, 231) In Hebrews, one of the late epistles written in Pauline tradition but not by Paul himself, we finally get what is often held as *the* definition of faith: "Now faith is being sure of what we hope for, being convinced of what we do not see." (Hebrews 11,1).

The latter definition remains remarkable for its subjective perspective and abstract form, i.e. for not mentioning the dogmatic contents like in Paul, John or in the Acts of the Apostles. Within Protestant tradition this difference between a subjective notion of trust and a dogmatic one focusing on its objective contents has again and again lead to harsh debates: Most prominently between Martin Luther (1483-1546) and Philipp Melanchthon (1497-1560) although both understood trust as an underlying subject of faith. Luther on the one hand argued for the indifference of the act of believing (*fides qua creditur*) and the contents of this belief (*fides quae creditur*) as both were given by God without expectance of any favour in return. In accepting this gift unconditionally man could reach *certainty* (NB: strictly distinct of security!) of salvation. Melanchthon, on the other hand, tended to reflect and analyse this existential faith in God's affection, asking how it could be realized by man. His answer – in *Loci communes* from 1521 – was threefold: In order to really understand God's love and gift we first need knowledge (*notitia*) of God and of the Gospel. For neither would we know about God naturally, nor would pure rational information suffice; in the first case we would not recognize God as he is revealed in the Bible, in the second case we would not necessarily become believers. So in second place comes assent (*assensus*) to the Christian gospel; we have to agree to God's Word, we need to be inspired. God's Word, however, contains gospel as well as law and threat. In order not to fall into despair facing law, we need – in third place – trust (*fiducia*) in God, in the gospel and God's promise. So in sum, faith in God is mainly trust in God's mercy, which is promised in Christ and even supported by occasional divine signs (Melanchthon 1997, 207-219).

What at first sight seems to have nothing at all to do with biobanking, is in fact crucial for our debates on trust today: On the one side we learn that trust is a deeply societal, interpersonal and/or intersubjective phenomenon. It is essential for social interaction, especially when conditions of interaction are unclear, when they raise doubt about the other's reliability. Herein is also implied that we need clear signals whether the other is trustworthy or not; those whom we shall trust in, have to prove themselves of being trustworthy. On the other side it becomes clear that trust, though it certainly needs thorough knowledge of the issue at hand, won't arise upon plain information. The latter might leave us better informed f. ex. about the main idea or certain processes in biobanking, but it would also leave us in knowing distance, it would not make it "our" project. In order to feel attached and committed to the issue at hand we have to consent to it as well, for only then will we be able to trust its promises for a better future. For both sides, however, it becomes clear that in trusting we admit a certain level of humility, even helplessness which has to be carefully transformed into confidence and thus trust again. The latter aspect is crucial, however, be it for ethical debates as the one on biobanking or in case of severe illness: Without any promising perspectives, trust will not be gained. If there seems to be nothing but decline and death in front of us, we will not be able to recollect our strength, take responsibility and act confidently again.

4.2 Enlightenment: The Parting of the Ways

In the course of history, to which we will now return, this close bond between confidence and responsibility has become crucial for the whole idea of trust. Although we have to note the astonishing fact that a detailed history of the concept of trust has not been written so far, we are at least able to reconstruct (f. ex. by using the index of the famous German encyclopedia "Geschichtliche Grundbegriffe", edited by R. Koselleck, where the appropriate article, however, is missing) a significant use of the term between 1650 and 1850, the very eventful time of Enlightenment and modern age, leading to the big cultural crisis in the middle of the 19th century: *On the one hand 'trust' is used in the political and moral context* for issues of state or nation, governmental affairs, questions of authority, liberalism or balance of powers. One of the most famous positions in this context is of course that of Thomas Hobbes who in "De Cive" of 1651 presented his idea of man's natural state as state of general mistrust, with competition, diffidence and glory-seeking as sources of conflict. Already here the problem of trust arises as a problem of social complexity, of general conditions of action and of long-term preservation of (social) trustworthiness which Hobbes sought to solve by theory of contractual law (cf. Hobbes 1983, 41-50; I/1-15). Different to this contractual line of thought in the English tradition based on Hobbes is the direction the "Deutsche Enzyklopädie" of 1779 chose. Here, the context is also a political one, namely the question in which way authorities in power should enact politics. The answer, however, comes closer to what we had read before in Aristotle, describing how *polis* worked. For here, too, it is without question that political settings usually are conflict-ridden so that authorities may uphold the law even by use of force. But the obedience gained thereby was useless, especially for the realization of governmental ends, namely to show people how to attain and enjoy prosperity (Rabe 1972, 395). Hence, authorities should instead act in such a way that people would understand the rational arguments underlying their programs and thus share them without further conflicts. For by acting upon rational and transparent arguments authorities would generate love and trust.[1] *On the other hand the notion of trust has become part of the concept of nature*: In modernity nature was generally understood in close similarity to the Aristotelian understanding: Nature is what is innate to man, what is defined by *autopoiesis* and is thus the very essence of objects. Leibniz, Kant and others were quite convinced that nature – as God's creation – was based on general and necessary principles waiting for man to be understood. Hence, in scientific research man sought to understand "the very nature of things" and therefore, as Kant held in the "Critique of Pure Reason", pressed nature to answer the questions posed by reason as it was reason stipulating the laws of nature

---

[1] „Es ist daher die größte Weisheit der Obrigkeit diese, daß sie, um die wirksamste Autorität zu erhalten, sich bei dem Volke immer mehr Liebe und Vertrauen erwerbe und alle ihre Anordnungen und Einrichtungen zu wahren Wohltaten für die Untergebenen mache." (Dt. Encyclopädie, vol 2, 609, cited in: Rabe 1972, 395).

(cf. Kant 1983, 23; KrV, B XIII). Yet, Kant had to admit in the "Critique of Judgement", reason could never fully understand the connection between general and particular laws of nature and had thus to suppose "the great whole" of nature, regarding it "as if" it were an organic system (cf. Kant 1983, 483-488; KU § 65). The more natural sciences, esp. chemistry and biology, took over from mechanical physics, the more the idea of nature as an organic system, as a dynamic process of oscillation became prominent as we can see in Romanticism and especially in Schelling's philosophy of nature (cf. "Erster Entwurf eines Systems der Naturphilosophie" [1799] or „System des transzendentalen Idealismus" [1800]). It is important to note the close connection to the increasing criticism of the hitherto idealized bourgeois principle, stressing freedom of the individual, reasonableness and utility. Romantic thinking opposed this by referring to the great whole, integrating everything individual. So whereas bourgeois thinking sought to break the spell of nature, romanticism established nature as opposing force to human reason. Consequently the idea of *natura naturata* (gaining products of nature) was replaced by that of *natura naturans* (nature producing) – even though in Schelling we find the idea that ultimately nature would be fulfilled nowhere else but in human consciousness. In the 19th century Hermann v. Helmholtz understood the reciprocal action of natural forces as one all-encompassing process; inherent natural laws could be proven by physical and chemical methods only. In his trust in the conformity of natural laws, however, the scientist would not only come to understand the conformity of natural law in the world, but also be able to subjugate nature.[2] The interesting move here is that the idea of "the very nature of things" was not only used in scientific thought but also in juridical contexts which might help to explain the further development. But in order to understand this process we have to be aware of the immense impact on the *power of control attributed to the idea of trust*, be it as a basic element of intersubjective and social interaction or as the key to the supremacy of nature – which at least in parts is due to its origin in Protestant tradition.

At first sight, however, both lines of thought seem to remain apart. For the period in the 17th and 18th centuries, which we first call Enlightenment and then the modern age, is the time of establishing bourgeois society claiming autonomy from all so far unquestionable authorities such as the Bible or the absolute power of the sovereign. The formerly self-evident unity of religion and politics had to be deconstructed in order to establish civil law. So it is in this context that the intellectuals of the time strictly rejected any biblical or theological ideas as normative sources of theoretical or practical reason. It was especially the philosophy of Kant that enforced the idea of the subject's autonomy making him famous for his strong notion of responsibility. It is less known, however, that this very idea of responsibility in Kant is based on a fascinating concept of trust, bound to the metaphysical

---

[2] „Mit seinem ‚Vertrauen auf die Gesetzlichkeit der Naturerscheinungen' werde der Forscher nicht nur ‚die Gesetzmäßigkeit in der Welt' begreifen lernen, sondern auch imstande sein, die Natur sich zu unterwerfen" (Schipperes 1978, 240).

notion of nature described above. Despite the fact that Kant stressed subjective autonomy, he remained deeply suspicious of human nature which he considered far too prone to emotions, passions and desire to fulfil the demands of pure reason. If, however, human nature in itself was a fragile instrument of reason's high ends, then his optimistic confidence in human subjectivity and responsibility had to be found somewhere else. And indeed, in his late "Critique of Judgement" of 1790, he expressed the idea that human life would fit nature as it was shaped according to the well-organized basic order of nature, designed by some higher intellect. Still, be it in his "Critique of Pure Reason" (1781), "Metaphysical Groundworks" (1785), or in "Critique of Practical Reason" (1788), normative ideas, he said, could only be found and formulated by reason itself so that whatever religion might contribute to our understanding of life, it had to remain "Within the Boundaries of Pure Reason" (1793) as the title of his last work suggested.

Hence, despite the fact that the theological tradition was harshly criticised and any explicit references to the biblical tradition and Christian faith were avoided, basic ideas of this tradition were nevertheless maintained. Kant, like most other intellectuals of the time, was raised in a pious family; he thoroughly knew the dogmatic tradition and was deeply influenced by it. Notions such as modesty of reason being aware of its boundaries, will being prone to temptation, subjective autonomy and responsibility being dependent on regulative ideas (such as God, freedom and immortality of the soul) or the idea of a higher order of nature have become popular and have not only been central ideas of Kant's philosophy but have shaped modern world – even though they have necessarily been transformed, f. ex. by the power of or at least by the desire for control mentioned above. Furthermore, and this is indeed remarkable for all further development, these genuine theological ideas are no longer recognized and/or addressed as such. Instead we have two different lines of thought, the former theological debate which is lead under the headline of Christian faith and belief, taking trust but as an underlying feature of faith, just as in Luther, as well as the non-theological debates in other disciplines which refer to trust as a topic of its own. Until today those two lines of discourse have remained apart to such an extent, that the theological debates seem to have been completely unaware of the increasing interest in the notion of trust, whereas the non-theological debates seem to have forgotten that extensive work had already been done on this topic long before. Evidently enough, this paper hopes to show in which way both lines could profit from each other – so let us now turn to prominent modern theories on trust.

4.3 Modern Theories: Social Interaction and the Law of Nature

It is no surprise to see that the two lines of thought sketched above, trust in the context of politics, juridical and social issues as well as trust in nature, have been of substantial influence until today and it is fascinating to see how in the issue of biobanking both aspects meet. The first issue has been taken up especially in the social sciences which have constantly been addressing the problem of political and economic complexity. Georg Simmel (1858-1919) f. ex. argued that modern society was based on the idea of credit and thus on mutual commitments, performed either in direct social interaction among trustworthy people (microlevel), objectified trust between professional partners (mesolevel) or in symbolic interaction like money for which no personal trust was required (macrolevel). Parallel to these three levels of trust, Simmel also distinguished between trust in the sense of general faith in seemingly self-evident processes (a), trust as a type of knowledge, knowing something but not being certain and thus placing trust in another person (b), and finally trust as feeling, placing trust in somebody else without even referring to the question of knowing or not in the sense of "meta-theoretical faith" (c), which, Simmel assumed, would occur in pure form probably only in religious contexts (cf. Endress 2002, 13-17). The idea of different forms of trust on micro-, meso- and macrolevel is also part of Niklas Luhmann's (1927-1998) famous study on trust of 1968, even though he is mainly following Talcott Parsons. The main idea of his functionalist theory is that trust allows us to reduce social complexity (cf. Luhmann 2000, 5), among other reasons because it substitutes knowledge when the latter is not to be gained (cf. Luhmann 2000, 31). Social life, Luhmann argues, is far too complex for us to fully understand, it offers far more possibilities and chances than we could ever dream to use, so there will always be a point where we simply have to act upon trust. According to Luhmann former societies were stable and rather simple systems, built on religious foundations and providing an atmosphere of trustworthiness (i.e. we are familiar with things and events, life itself need not to be questioned) whereas modern societies are based on decision processes and constitutive acts leading to an increase in complexity, especially as they are organized by general media of communication like money, truth or power. On the one hand such general media would reduce complexity but on the other hand they have to be implemented, they have to be put to the test, they first have to be trusted in order to become trustworthy and work in case of emergency (cf. Luhmann 2002, 74). Personal trust thus changes into system trust, i.e. the conviction that systems have proven a success (cf. Luhmann 2000, 90). Only if societies manage to establish and strengthen trust in their systems will they become stable again and thus be able to flourish. A prominent example for Luhmann's idea of system trust can be seen in the current financial crisis: Despite all warnings, trust in systems based on subprime loans was upheld until the final crash, leading to an immense loss of trust, destabilizing the whole financial system which will now be rather difficult to re-establish.

It is interesting to see how prominent the notion of trust has become during the past few years, be it in academic research or in public, spreading to newspapers and magazines. Many theories seem to be strongly influenced by rationalistic approaches, seeking to find out whether at all and if yes, when, where and in whom it is reasonable to trust. In rationalistic approaches such as rational-choice-theory, economic theories, game-theory or signalling theory (all excellently presented in

Möllering 2006, 15-43) trust is part of social-action-theory, based on purposive actors, carefully estimating the risk of trusting others in order to get rational decisions on the issue at hand. Alternative modes of action, calculation of possible obstacles, visualization of different possible outcomes, balancing of self-interest, risk (actors and the possibility of misplaced trust) are carefully estimated by any trustor.

> The [rational, CR] paradigm", Möllering states on rational-choice-theories, "explains trust as a rational result of self-interested actor's perceptions of another actor's trustworthiness. Those perceptions are unanimously seen as imperfect estimates of (a probability of) trustworthiness, which makes acting on trust risky for the actor. Variations within this paradigm concern questions such as what kind of information the actor needs to consider, whether trust is a cognitive or behavioural category and how serious the conceptual limitations of the paradigm are. The last point enables certain conceptual extensions: while irrationality or non-rationality are recognized as a possibility by all proponents, only some of them attempt to bring them into a model using constructs such as norms, culture and predisposition. Rational choice theory is a social theory perspective based on an economic logic (Möllering 2006, 24).

Economic theory, however, has its focus in averting negative outcomes which makes calculativeness the main mechanism of behaviour, aiming at the economization of transaction costs (cf. Möllering 2006, 26). In addition, principal-agent-theories or game-theories complement a discourse which in fact, Möllering argues, is a lot more interested in models of "cooperation" but not in trust, "equating the two terms quite liberally" (Möllering 2006, 41). Yet, he also concedes that this has already been seen, e.g. in Williamson 1993 or James 2002 who differentiate between trust and cooperation. Here, trust is understood as an "independent exogenous parameter that explains why cooperation occurs when the theory predicts that it should not" (James 2002, in: Möllering 2006, 41). Hence, economic concepts often "require a positive net expected value from trusting", they distinguish "between rational acts of cooperation and non-rational acts of trust", leading to the "paradox", that in the end economic theories tend "to explain trust away", as Möllering concludes (Möllering 2006, 43).

The second line, referring to trust in the conformity of natural laws which enables us to understand the conformity of natural law in the world and thus subjugate nature, can best be illustrated by the positions of Alfred Schütz (1899-1959) and Erikson. Schütz (to the following cf. Endress 2002, 17-21) developed his theory following Edmund Husserl's phenomenology, especially in the latter's idea that human lifeworld ("Lebenswelt") is experienced as something self-evident, as something un-questioned, as something we naturally act in because of its familiarity – at least as long as this basic familiarity is not disturbed. Within this familiar context we are able to experience new things and to enlarge our knowledge. Here Schütz follows W. James who distinguished between (objective) knowledge of acquaintance (i.e. knowing *that* something is the case) and the (deeper) knowledge about *how* things happen. Both levels shape our perceptions and actions in the world, but as the latter refers to its basic structures, it is here that our natural trust in our lifeworld is confirmed. Its main objectives are a) temporal dimensions (past, present, future) and their subjective correlates (memory, fulfilment, expectation), b) social structures (fellow human beings, ancestors, descendants), and c) dimensions of action (present, restorable, potential). Given a high level of familiarity in these respects, human beings would simply rely on the world, on the reliability of our experiences of the world and on our ability to act in this world as something undoubtedly and constantly given. This idea of a kind of "natural trust" was later taken up by many authors, among them Erik H. Erikson (1902-1994) as the perhaps most famous exponent. His notion of "basic trust" became highly influential in developmental psychology, pedagogic, philosophy and other disciplines. His basic idea is usually repeated in the following way: Within the first developmental steps of a child, starting in the womb, crucial conditions of his/her further psychological development are laid, namely the ability to either meet the world in "basic trust" or in constant mistrust. In "eight stages of man", reaching from "trust versus basic mistrust" to "ego integrity versus despair" the ego would gradually gain a feeling of integrity and identity (Erikson 1950, 219ff.). So educational theory has been busy to explain how parents could best take care of this risky task by providing an atmosphere as safe and trustful as possible. Though there is nothing wrong with this reading of Erikson, it remains unconscious of its background: In his 1950 study "Childhood and Society" Erikson in fact intended a "psychoanalytic book on the relation of the ego to society" (Erikson 1950, 11f.) by analysing different cultural and educational settings, ranging from anxiety in young children, apathy in American Indians, confusion in veterans of war or arrogance in young Nazis. The idea which allowed him to combine such a broad and at first sight almost accidental range of different settings was the following: all of them were due to certain conflicts between ego and society. In psychoanalytic perspective, Erikson says, "we learn that a neurotic person, no matter where and how and why he feels sick, is crippled at the core, no matter what you call that ordered or ordering core. He may not become exposed to the final loneliness of death, but he experiences that numbing loneliness, that isolation and disorganization of experience, which we call neurotic anxiety." (Erikson 1951, 20) In seeking to reconstruct the history of these individual conflicts by interpreting medical data in their relation to past experiences, the psychoanalyst would study "psychological evolution through the analysis of the individual. At the same time it [psychoanalysis, CR] throws light on the fact that the history of humanity is a gigantic metabolism of individual life cycles." (Erikson 1951, 12) So the idea behind the notion of "basic trust" is not simply a question of properly raising a child in a supporting way, but: „it is first of all important to realize that in the sequence of these habits [which the child gradually develops in the earliest stages, CR] the healthy child, if halfway properly| guided, merely obeys and on the whole can be trusted to obey inner laws of development, namely those laws which in his prenatal period had formed one organ after another and which now create a succession of potentialities for significant interaction with those around him. While such interaction varies widely from culture to culture, in ways to be indicated presently, proper rate and proper sequence remain critical factors in these successive manifestations." (Erikson 1951, 62f.) The power of this background idea, however, is best to be seen in Erikson's later study "Identity and the Life Cycle" from 1959 where he further explains the developmental stages. The crucial point here is that Erikson calls on the principle of *epigenesis*, stating "that anything that grows has a *ground plan*, and that out of this ground plan the *parts* arise, each part having its *time* of special ascendancy, until all parts have arisen to form a *functioning whole*." (Erikson 1968, 52) Hence, even though "a reasonable amount of guidance" is needed, "the healthy child [...] can be trusted to obey inner laws of development, laws which create a *succession of potentialities for significant interaction* with those who tend him. [...] Personality can be said to develop according to steps predetermined in the human organism's readiness to be driven toward, to be aware of, and to interact with, a widening social radius, beginning with the dim image of the mother and ending with mankind, or at any rate that segment of mankind which 'counts' in the particular individual's life." (Erikson 1968, 52) The message behind it is simple: Nature provides life and human life in it with all it needs in order to successfully develop. Any physical or neurotic disorder is therefore due to some sort of disturbance of what originally would be a sound process. Different to the notion of confidence, Erikson argues against Therese Benedek, the notion of trust would express "more naïveté and mutuality" and it would imply that one has learned "to rely on the sameness and continuity of the outer providers but also that one may trust oneself and the capacity of one's own organs to cope with urges; that one is able to consider oneself trustworthy enough so that the providers will not need to be on guard or to leave." (Erikson 1968, 61) It is evident how close Erikson is to the antique notion of nature, especially to Aristotle: Nature is what is innate to man, natural objects bear within them the beginning of development and standstill/stagnation, they evolve in the process of *epigenesis* – an idea we find but slightly changed in the romantic notion of *natura naturans* described above.

5 Conclusion: Trust – *Grenzerfahrung* Transforming Anxiety into Confidence

So in the end we seem to have returned to the antique notions we started with. What then, have we learned from this survey of historical developments and systematic reflections on the relation between trust and responsibility, covering risk, safety and security? What have we learned for the ongoing debates in biobanking? Does it indeed all come down to the question whether we trust in biobanking or not? Above all it should have become obvious that trust is not only a phenomenon that appears in multiple perspectives but that it also comes into play whenever questions of risk, safety, security and responsibility are posed. It is therefore no surprise that in public debates biobanking is, among other things, considered a matter of trust. But in order to understand the different and sometimes concealed levels of the debate, we have to clearly distinguish between the different dimensions implied.

The first and perhaps most obvious aspect is that in biobanking technical, medical and scientific issues are at stake which demand to be taken care of. Technical safety of laboratories, production and storage, reliability of administrative and financial procedures or qualification of staff are and will be issues to constantly improve on in order to maintain the *safety* of the whole project. Hence, it had definitely been necessary, as Catherine Whitbeck argued in 1995, to discuss biobanking in respects of technical improvement as well as improvement in quantitative and qualitative procedures. But, she said, after the solution of most technical problems new difficulties would arise, f. ex. concerning the reliability of researchers:

> "We have recently reached a watershed in the research community's consideration of the ethics of research. The way is now open for a more nuanced discussion than the one of the last decade which was dominated by attention to legal and quasi-legal procedures for handling misconduct. The new discussion of ethical issues focused on trustworthiness takes us beyond consideration of conduct that is straightforwardly permitted, forbidden or required, to consideration of criteria for the responsible as contrasted with negligent or reckless behavior." (Whitbeck 1995b, 403).

Even though huge efforts have been undertaken to improve what in 1995 has certainly been a crucial point, latest scandals have shown that this part of the agenda is definitely in need of more effort.

Yet, this would still not be enough in order to reach the point Anne Cambon-Thomsen correctly stated almost ten years after Whitbeck in 2004, when she noted a peculiar shift in public attendance or, to use a better word, attention towards biobanks, namely from perceiving them as clinical or academic research tools to subjects of *societal* debate. As such biobanking almost inevitably turns into a subject of trust for throughout history the latter has been part of intersubjective, social and political affairs, especially if legal aspects are implied. So in addition to the first level of debate, mainly aiming at reducing (technical, medical, administrative etc.) risk in order to guarantee safe procedures, the second level points to possible intersubjective and social causes of trouble which are manifold. As soon as biobanking had become a successful tool of clinical and/or academic research, it became public simply because of the sheer amount of people being involved, ranging from scientists to politicians and financiers to patients. In this context biobanking is no longer an issue for experts but is also relevant for people who have no idea about the project and might therefore react reservedly and carefully, especially when they are confronted with the topic in a situation of (severe) illness.

It is interesting to note that it is this part of the argument, which is best captured in the concept of trust itself, at least in the English-speaking debates. According to the "Encyclopedia Britannica", the concept of *trust* is mainly a term of jurisdiction and business, suitable for defining cases of property between definite persons (cf. Encyclopædia Britannica 2009b). Related links are then "trust in business, "trust companies" i.e. legal corporations and foundations (e.g. Getty Trust), "charitable trust", "investment trust" and "trust territory". There is but one link referring to "religious philosophy", namely a short reference to Martin Buber's study "Zwei Glaubensweisen" from 1950. So the (former) American President, saying: "In God we trust", might very well have had a juridical relation in mind. Even related notions such as "confidence", "belief" or "faith" are influenced by this immense power of juridical thought, although they also come close to the religious tradition sketched above. For even the notion of confidence, which – from a German-speaking point of view – we might assume to refer more to the subjective inner mode in the sense of familiarity, of confidentially approaching others, is in the "Encyclopaedia Britannica" reserved for "confidence games (swindling operations)", with related links pointing to "confidence interval (statistics)", "vote of confidence (government)" or "professional confidence (privileged communication in law)". The term "belief", then, is defined as a "mental attitude of acceptance or assent toward a proposition without the full intellectual knowledge required to guarantee its truth." (Encyclopædia Britannica, 2009a). Only in the articles on the concept of faith this line seems to be changed as the main article here is clearly dedicated to religious faith as an "inner attitude, conviction, or trust relating man to a supreme God or ultimate salvation. In religious traditions stressing divine grace, it is the inner certainty or attitude of love granted by God himself. In Christian theology, faith is the divinely inspired human response to God's historical revelation through Jesus Christ and, consequently, is of crucial significance" (Encyclopædia Britannica, 2009c). This matches what we said about the Christian tradition, yet, even here we find the notion of "good faith (law)": "Perhaps the most important principle of international law is that of good faith. It governs the creation and performance of legal obligations and is the foundation of treaty law" (Encyclopædia Britannica, 2009c).

Nevertheless, neither referring to technical and medical problems nor referring to social and juridical aspects involved in the relation between biobanking and trust would suffice. For the short reflections on the notions of risk, safety/security and responsibility showed a problematic picture· What used to be accounted for in terms of objective risk, provided security and individual responsibility has merged into a conglomeration of medial, technical, scientific, economical, social, moral, international and intercultural aspects. So even though we still expect many aspects of biobanking to be dealt with in terms of reducing risk and providing safety we are increasingly aware of and concerned about the untenable nature of this expectation – and it is exactly that moment which calls upon the notion of trust. What we need therefore are studies not only on scientific progress but also on source and background of experiences of uneasiness, feelings of uncertainty, anxiety and even threat.

The antique tradition already required trust for social interaction especially when conditions of interaction were unclear and potentially dangerous. Generally, trusting somebody meant to admit a certain level of helplessness which carefully had to be transformed into confidence. Promising perspectives were needed as otherwise one would not be able to recollect one's strength, take responsibility and act confidently again. Even though in and through Enlightenment we find a loss of explicit theological traditions, the ideas behind them, if it came to 'trust', never subsided. For only superficially the separation of ways between the antique philosophical and then theological positions and the younger rational, sociological, economical or psychological approaches seems to be self-evident. On closer inspection all of them not only mirror but even refer to the crucial fact that in the end – be it in outspoken argument or in hidden fear – questions of risk, safety and security are not to be solved or eliminated. Instead most of them in one way or another express fear of losing security and control. What had been evident for Luther and his successors, namely that in trusting we admit a certain level of helplessness, refrain from the temptation of replacing trust by security and thus gain certainty and confidence to consent, is in fact also figuring here: Simmel and Luhmann take society as being based on credit and mutual commitment, demanding general faith, knowledge and a special kind of (religious) feeling in order to reduce social complexity. Rational approaches, on the other hand, ask when, where and whom to trust. They seek to answer accordingly to economic logics, turning trust into a question of calculativeness. What they do not realize, however, is that this idea of calculativeness cannot work as we have learned from the history of risk, safety/security and responsibility. Schütz, last but not least, argued for the general reliability of "Lebenswelt" in the sense of basic familiarity, which according to Erikson works fine – but only as long as it is not disturbed. So no matter how different these theories are, all of them indicate but one thing: the need for security in its broadest sense, reacting to the hidden anxiety of loosing control. The crucial point, however, lies in the question whether and in which way we admit of this helplessness or not.[3]

"Trust" therefore has to be understood as something which demands knowledge and consent while still taking experiences of uneasiness and fear seriously and thus respecting the basic need for security. In order to gain trust, biobanking indeed has to prove itself trustworthy. This means, to offer sufficient and understandable information, be it on paper or internet, and to improve on all communication upon the matter, be it in doctors' surgeries, hospitals or research projects. Here, however, one tiny little detail has to be taken seriously: While prominent theories, esp. in rationalistic lines of thought, seek to stress the importance of knowledge as means of correctly assessing problems at hand and thus gaining security because ignorance might lead to false decisions, plain information will never suffice. Instead it might be useful to refer to and reintroduce the threefold Protestant notion of *notitia*, *assensus* and *fiducia* presented above as such an approach would help to first and better sort out the different perspectives of the prob-

---

[3] In some ways this comes close to what Annette Baier (1994, 2001 et al.) said about trust as accepted vulnerability or as leaving one's concerns to somebody else, but as far as I can see she developed this mainly in discussing Hume, not in following the philosophical and theological line presented here.

lem at hand. For plain information would not suffice to soothe feelings of uncertainty, anxiety and even threat. As they might heavily influence people's willingness to collaborate with academic research, consultation processes have to explicitly address such experiences in order to find their sources. For only if patients fully agree with and consent for the project at hand will they provide reliable information. So despite the fact that from modern age onwards, trust was no longer under debate as a theological question, this tradition might offer new perspectives for present debates.

Yet, referring to this threefold notion can still not explain sufficiently why the debates are lead in such an agitated way. I would therefore like to suggest to proceed one step further in order to better understand the notion of trust or *fiducia* itself. For even though one could be perfectly informed about things, feel very well-advised and explicitly asked to consent, one might still not be able to trust. This can f. ex. be seen in case of severe illness: Even though one might possess all information possible on changes and limitations of certain therapies, even though doctors seem to be reliable and caring, even though final decisions on therapies are up to oneself, one might still be caught by sheer fear of death. So despite the fact that we so far have reconstructed many features of the notion of trust I think further examination is needed to understand the basic dimension of anxiety implied. In some way, trust – in its emphatic sense – seems to be bound to *Grenzerfahrungen*: fathomless anxiety seems to engulf everything reliable, crucially revealing our limitations so that all stability seems to be lost. It is evident that such experiences hit us in utmost intimacy, breaking all confidence and thus our ability to act confidently. Yet, at the same time it is in trusting that we are able to generate confidence and responsible action again. So in order to understand both, trust as well as responsibility, we will have to better understand the transformations in and through *Grenzerfahrungen*.

References

Aristotle (1985) Nikomachische Ethik. PhB, vol. 5. Felix Meiner, Hamburg

Aristotle (1990) Politik. PhB, vol. 7. 4th ed. Felix Meiner, Hamburg

Baier A (1994) Moral Prejudices. Essays on Ethics. Harvard Univ. Press, Cambridge et al

Baier A (2001) Vertrauen und seine Grenzen. In: Hartmann M, Offe C (eds.) Vertrauen. Die Grundlage des sozialen Zusammenhalts. Campus, Frankfurt a. M.

Bayertz K (Ed.) (1995) Verantwortung: Prinzip oder Problem? Wissenschaftliche Buchgesellschaft, Darmstadt

Blackman EC (1982) Faith, faithfulness. In: Buttrick GA, Crim KR (eds) The Interpreter's Dictionary of the Bible Vol 2, 13th ed. Abingdon, New York, Nashville: 222-234

Cambon-Thomsen A (2004) The social and ethical issues of post-genomic human biobanks. In: Nature Nov. 2004: 866-873

Conze E (2005) Sicherheit als Kultur. Überlegungen zu einer "modernen Politikgeschichte" der Bundesrepublik Deutschland. In: VfZ 53: 357-380.

Encyclopædia Britannica Online (2009a) Belief. www.britannica.com/EBchecked/topic/59442/belief. Accessed 18 Jan 2009

Encyclopædia Britannica Online (2009b) Trust. www.britannica.com/EBchecked/topic/607348/trust. Accessed 18 Jan 2009

Encyclopædia Britannica Online (2009c) Faith. www.britannica.com/EBchecked/topic/200515/faith. Accessed 22 Jan 2009

Endress M (2002) Vertrauen. transcript Verlag, Bielefeld

Erikson EH (1951) Childhood and Society. W.W. Norton & Company, New York

Erikson EH (1968) Identity and the Life Cycle. Selected Papers. With a historical Introduction by David Rapaport. 3rd ed. International University Press, New York

Gloyna T (2001) Vertrauen. In: Ritter J, Gründer K, Gabriel G (eds) Historisches Wörterbuch der Philosophie vol. 11. Wissenschaftliche Buchgesellschaft, Darmstadt: 986-990

Hansson SO (2007) Risk. In: Zalta EN (ed) The Stanford Encyclopedia of Philosophy http://plato.stanford.edu/archives/win2008/entries/risk/

Harris J (2005) Scientific research is a moral duty. In: J Med Eth 31: 242-248

Hobbes T (1983) De Cive (1651): The English version. A crit. ed. by Howard Warrender, RP 1983. Clarendon Press, Oxford

Jepsen A (1973) amn. In: Botterweck GJ, Ringgren H (eds) Theologisches Wörterbuch zum Alten Testament vol. 1. Kohlhammer, Stuttgart: 314-348

Jonas H (1984) The imperative of responsibility. In search for an ethics for the technological age. University of Chicago Press, Chicago

Kant I (1983) Kritik der reinen Vernunft (1781/86). In: Weischedel W (ed) Kant, Werke in zehn Bänden, vol 3. Wissenschaftliche Buchgesellschaft, Darmstadt

Kant I (1983) Kritik der Urteilskraft (1790). In: Weischedel W (ed) Kant, Werke in zehn Bänden, vol 8. Wissenschaftliche Buchgesellschaft, Darmstadt

Knight FH (1964) Risk, Uncertainty and Profit. Kelley, New York

Luhmann N (2000) Vertrauen. 4th ed. Lucius & Lucius, Stuttgart

Makropoulos M (1995) Sicherheit. In: Ritter J, Gründer K Historisches Wörterbuch der Philosophie vol 9. Wissenschaftliche Buchgesellschaft, Darmstadt, 745-750

Melanchthon P (1997) Loci communes 1521. 2nd ed, Gütersloher Verlagshaus, Gütersloh

Möllering G (2006) Trust: Reason, Routine, Reflexivity. Elsevier, Oxford, Amsterdam

Rabe H (1972) Autorität. In: Koselleck R (ed) Geschichtliche Grundbegriffe. vol. 1, Klett-Cotta, Stuttgat: 382-406

Rammstedt O (1992) Risiko. In: Ritter J, GründerK (eds) Historisches Wörterbuch der Philosophie vol.8. Wissenschaftliche Buchgesellschaft, Darmstadt: 1045-1050

Schipperes H (1978) Natur. In: Koselleck R (ed) Geschichtliche Grundbegriffe. vol. 4, Klett-Cotta, Stuttgart: 215-244

Whitbeck C (1995a) Trustworthy Research. Editorial Introduction. Sci Eng Eth 1: 322-328

Whitbeck C (1995b) Trust and Trustworthiness in Research. Sci Eng Eth 1: 403-16

Zweig S (1982) Die Welt von gestern. Erinnerungen eines Europäers. 2nd ed, S. Fischer, Frankfurt a.M

Ethical Issues

Which Duty First?

An Ethical Scheme on the Conflict Between Respect for Autonomy and Common Welfare in Order to Prepare the Moral Grounds for Trust

Peter Dabrock

Abstract Biobanks will only be established successfully and run sustainably if the principles, procedures of governance, management, control and participation are built solidly on trust. Against this backdrop, this paper proposes the idea that – subject to the development of some reasonable conditions – we cannot deny a certain moral obligation to participate in forms of biotechnological advances in genomics such as biobanks. In order to develop a sound argument, the theological and ethical approach that is taken in this contribution is sketched out. Thereby, a special (and not a general) account of moral obligations to participate in biobanking is defended. Such a scheme will, from its outset, take into account existing fears and anxieties but will serve not to assert these fears in general and therefore abolish the use of genomic-based measures, but to find a responsible way to deal with them. Therefore, the concepts of "genetic exceptionalism" and "persons genetically at risk", which are prevalent in the debates about genomic research in general and biobanking in particular, are rejected. Dealing with the question whether there are obligations to participate in biobank research and which criteria might be relevant to determine the conditions, which make it legitimate to assume and to gradualize such obligations, an ethical model for balancing respect for autonomy and common welfare is developed, presented and defended.

1 Introduction

Biobanks will only be established successfully and run sustainably if the principles, procedures of governance, management, control and participation are built solidly on trust. No such endeavour will function successfully without a commitment to engendering confidence in those who are thinking about donating samples or releasing data to a biobank. Anyone who attempts to run a biobank enterprise must be fully aware of the technical setting and socio-ethical benchmarks with regard to privacy, confidentiality, non-discrimination, non-stigmatisation, participation and benefit sharing. In addition, benchmarks of good governance as promoted by the UN Economic and Social Commission for Asia and the Pacific must be taken into account: those raised include consensus orientation, accountability, transparency, responsiveness, effectiveness and efficacy, equity and inclusiveness, following the rule of law and awareness of current and future needs of the society.[1] The OECD initiative for the preparation of guidelines for human biobanks and genetic research is a move in the same direction.[2] Based on principles such as the protection of human biological materials and data, access, custodianship, benefit-sharing and intellectual property, it seeks to provide a framework of best practices and ways of applying them through governance and management strategies, qualification and training methods and assessment techniques.

From a deontological approach in normative ethics, these standards are assumed and upheld, because everyone who is willing to contribute to a biobank enterprise must be recognized as a person – meaning an end in him or herself. If your preference is consequentialist ethical approaches or economically driven thoughts, you will probably also adhere strictly to the high standard of ensuring trust or flippantly risk the success of your labour. To break even is the crucial criterion for these approaches.

How to cope with public anxieties and fears concerning the establishment and management of biobanks is not the only topic of ethical concern in this field. As ethics cannot be identified with the task of law, it must be permissible to raise ethical questions which may have only a very indirect impact on immediately forthcoming political and juridical decision making. It may nevertheless be challenging to consider these thoughts far from immediate application if they affect a point of moral relevance. I am convinced that such a point is encountered in the consideration of the duty to contribute to endeavours such as biobanks. Concretely, I would like to propose the idea that subject to the development of some reasonable conditions, we cannot deny a certain moral obligation to participate in new forms of biotechnological advances in genomics such as biobanks. Although I have always supported the concepts of human dignity, self-determination and the entitlements derived from these ideas (Dabrock et al 2004), and have been aware that some approaches in the social sciences have alarmingly implicit or explicit tendencies to infringe the mentioned basic entitlements and capabilities indicative of human flourishing (Nussbaum 2000) I am nevertheless satisfied that we cannot deny, and must therefore support, the idea of some moral obligation to participate in certain types of genomic research. The whole argument depends, however, on how this obligation is understood – firstly in its justification and secondly in its practical outworking. It might be that a pragmatic implementation of a theoretical obligation to participate in biobanks departs profoundly from the ethical justifica-

---

[1] Cf. http://www.gdrc.org/u-gov/escap-governance.htm. It goes without saying: inescapably you get into conflict when different interpretations of these buzz words meet. For applying the concept of governance to the field of biobanks cf. Gottweis (in this book).

[2] While this paper is published the process is still running (cf. www.oecd.org/sti/biotechnology/hbgrd).

tion. Practical implementation of an idea may be achieved by use of incentives, whereas the concept of justification strives for obligations. From an ethical point of view, no offer of enticement may be considered legitimate without ethical warrant.

As a means of presenting the general idea and then a step-by-step scheme as to how to deal responsibly with this alleged moral obligation, I will start by sketching out my theological and ethical approach in form and content, and thus take the opportunity to test the truth of my claims. I will then provide reasons as to why I am convinced that an obligation to participate in genomic research is currently at stake. I do not intend to defend a very general concept of moral obligations, as wide ranging approaches often fail to be applied appropriately, but will confine myself to talking about a special scheme. Such scheme will from its outset take into account existing fears and anxieties but will serve not to assert these fears in general and therefore abolish the use of genomic-based measures, but to find a responsible way to deal with them. How this does or does not work will be the topic of the preliminary discussion and some subsequent parts of this paper.

2 The Approach of Theological Ethics

The notion of an obligation to participate in genomic research and measures will have to meet general standards of theological ethics. You will find a broad consensus in this area, from nearly all theologians to those who respect and defend the human rights tradition and recognize each human being as a bearer of human dignity. From the specific angle of protestant theology, the interpretation of this axiom must address two main traits when considering potential obligations for individuals in the field of genomics. Firstly, a protestant understanding of human dignity emphasises the liberty and freedom of each individual. This value is derived from a combination of the central theme of Christian liberty originating from Paul to Augustine and Luther with the core cultural feature of modern times and enlightenment. On another occasion I could or should talk about the theoretical chance, challenge and threat of this melting pot of European heritage in more detail. Here it is worth drawing one decisive conclusion from this axiom: obligations and prohibitions that do not obviously prevent actions that endanger the liberty of others must be substantiated. In other words, the burden of proof lies with those who claim the existence of an obligation. This holds true for individuals as well as for organisations in institutional settings.

Secondly, ecclesiastical and theological traditions are very sensitive to the fact that at this end of human life, self-determination is still and always under peril, especially for those who are named as groups of disadvantaged and vulnerable individuals and populations. Together with other philosophical approaches and social movements, the broad range of Christian traditions presents the idea of social justice as a means of really – not only formally – enabling each individual to life a live according to his or her own understanding of a good life, and to provide the opportunity of real participation in very diverse societal communications. Appropriate means and measures, corresponding to rights and liberties, must be provided to facilitate the inclusion of each individual in society and to give him or her a chance to shape societal life, especially education and literacy, to a level of sufficiency derived from this understanding of human dignity and communicative liberty (Nussbaum 2006, Powers and Fadden 2006).

3 Why Obligations Matter in the Field of Biobanking

Why do I uphold the assumption that, apart from these fundamental limitations upon asserting them at all, obligations should not be dismissed either generally or in the particular field of genomics and biobanking? Justification of such an assumption seems possible only under very specific conditions in form and content. At least two presuppositions are to be taken into account. First, I want to call attention to moral obligations, rather than legal ones. In contrast to approaches of ethics and political philosophy that either restrict obligations to the legal sphere or present them as strong obligations which stand the proof of universalization, I find it necessary to talk about obligations not only as "perfect obligations as they are understood by Kant" (Kant 1959, 421; Kant's term implies that it is prohibited to ignore the obligation)[3] but also as those that are imperfect, which are conditional, which are weak, which cannot be compelled, but are nevertheless praiseworthy and which are stronger than those that are only supererogatory (Heyd 2006).

Provision of support for those in trouble and not able to relieve themselves of their difficulties can be reckoned among this class of obligations, especially where there is no harm to fear from supporting the disadvantaged. Despite the fact that, apart from criminal failure to render assistance, the law may keep silent on such cases, from my point of view ethics cannot do so. On the contrary, moral and normative ethics derived from the above-mentioned axioms should promote sensibility and sensitivity to obligations if they are strongly connected with what are often called "natural duties", signifying their widest acceptance throughout different times and cultures. As they suggest themselves for all peoples in the world, regardless of culture, and seem evident at first sight, they have also been coined "prima-facie" obligations.[4] Among the cross-cultural norms observed nearly always and everywhere you find imperatives such as "Do not harm!" or "Do good to

---

[3] Cf. For an overview Johnson 2008: "Perfect duties come in the form 'One must never (or always) φ to the fullest extent possible in C', while imperfect duties, since they enjoin the pursuit of an end, come in the form 'One must sometimes and to some extent φ in C'."

[4] This does not rule out the formulation of more nuanced definitions – rather the opposite is the case: depending on the precise situation for which a given obligation is relevant, further modifications, also with regard to other rules, may become necessary.

those who are in need!" or: "Do more good for those who need it most!" The trouble starts when different interpretations of these demands clash with each other.

Applying these ideas to the field of social ethics and political philosophy, they may be reformulated as follows: in society there are obligations to be fulfilled, usually by the state, firstly not to infringe fundamental rights, for example that of academic freedom, and secondly, as mentioned above, to provide a sufficient level of conditional goods for each individual, for instance health, shelter, food etc. In concrete terms, rights and entitlements are to be guaranteed by society, primarily by the state, that is the organisation that makes the decisions in the political sphere of the society.[5]

The second presupposition that must be taken into account when talking about obligations from a liberty-oriented ethical perspective admonishes one to remain aware of the idea of the common good or the common welfare, understood not only in an utilitarian way, but in a way that is also generally accepted by the last developments of liberal political philosophy under the pressure of communitarianism: the idea of a thin good or a strong vague understanding of the good. This means that it is necessary to support some values and capabilities for the sustainable stabilization of society and to foster the situation of each individual to a decent level by improving research and education, not only by funding them but also by placing them at the top of the agenda, making them ends in themselves.

---

[5] In political philosophy, there is broad-ranging discussion about the ways in which the obligation to obey to the state and its institutions can be justified. This discourse cannot be presented here in detail and does not need to be considered at length for our purpose; take for example the question as to contributions to measures which guarantee welfare outside of the personal life of the payer. From my point of view, the following explanation should suffice to legitimate these obligations: human life in its individual and social aspects can only be successful if people respond to calls for help, except where in doing so they would expose themselves to hardly bearable risks. This has been referred to by Christopher Heath Wellman as "samaritanism". The ensuing reciprocal expectations, which are often established spontaneously on the interpersonal level, can be mapped and standardized on the social level. A discussion about this issue includes and combines the aspects of fairness, reciprocity, recognition and universalization. It is a universal insight that fairness implies that the desire for help in one's own emergency is linked to an obligation to help others who happen to be in a similar situation. This is independent of any individual experience with emergencies, because the mere thought of such a situation can be interpreted as a moral reason for a prima-facie obligation. Consequently, everyone who states that universalization presupposes a disregard for group-specific particularities, labours under a misapprehension ("agent-neutral approach"). This demand again contradicts wide-spread intuitions. As a matter of fact, under ceteris-paribus conditions and if only one person can be saved, many people would prefer to rescue a close relative instead of someone they would not know. Indeed, this restriction on prima-facie obligations can be universalized as well – it is allowed to give preferential treatment to a close relative or person under ceteribus-paribus conditions in an emergency. As a consequence, we can still universalize the prima-facie rule: you are expected to help people in case of an emergency. The discussion will therefore not focus on the rule itself, but rather on the question as to whether its preconditions are met in a specific situation: the need of help on the part of the potential patient and temporal, situational and social factors on the part of the actor or helper; cf. for an overview Dagger 2007.

Even at the level of general ethics, objections against this general approach can be raised. Obligations must be balanced against rights and entitlements and assessed with regard to the agent (her personal capacities and abilities, her societal or professional possibilities and options), her (symmetric or asymmetric) relation to the addressee (friendship, kindredship, professional relationship, socio-political circumstances), the external constraints (standard or exceptional situation) and anticipated consequences for the persons involved and society in general. All this cannot and should not be denied. Nevertheless, none of these objections are principal arguments against the consideration of obligations when the standards of a dignity based approach are taken seriously. But the mentioned objections call for specific justifications and differentiations when thinking about moral duties.

When pondering a probable obligation to undertake some form of genetic testing or to participate in genomic research like biobanks,[6] the question is whether or not the preconditions of obligation discussed earlier are at stake. In such circumstances – a possibility of genetic testing or contribution to biobanks by donation of samples or provision of personal data information – do we confront people in need? When talking about genomic research, do we meet a vitally important and imminent matter, for instance of common welfare, which does not admit delay?

Of course, there are only a few things of absolutely crucial emergency. I am convinced however that despite all the risks still to be taken into consideration such as stigmatisation, discrimination, the build-up of biosocieties (Rabinow 1996) and the concept of genetic risk (Rose 2007, 106-130) including exclusion and infringement of recognition of the other as an Other, the general course of genome-based medicine contains a promise that is of major importance. The importance is attributable to the non-neglectable idea of a more, not fully personalized and tailored medicine, the idea of a better stratified medicine that could be carried into effect not only for those in the developed, but also for those in the developing world (World Health Organisation 2002, 79-105). Anxieties aside, I can imagine scenarios in which it would seem to be a violation of moral obligation not to recommend the usage of genetic tests. With these realistic scenarios in mind, the following question comes to the fore: might a development that will probably, more or less, sooner or later become a reality give rise to an obligation to participate in it? I do not suggest a carte blanche to undertake participation in all kinds of genomic research in general and biobank endeavours in particular. I simply intend to introduce and provide a differentiated scheme of obligation. But to consider such a scheme at all presupposes that such developments may in general be considered

---

[6] The objection covering genetic testing and genomic research with one approach would generalize special problems inappropriately may be raised. Genetic testing is a mean in order to cope with personal health status seems to be a completely different issue than participating in a research endeavour that looks for transpersonal health outcomes like surveys of genetic epidemiology. No one would argue that point. Nevertheless, both actions do not only raise ethical questions how to protect each individual but ask for the social impact the personal responses have on the society in general. It is only this point I would like to focus on without denying the core relevance of the frequently debated topics of protecting the individual.

worthy. There are at least two reasons for this general assumption: first the relevance of basic and biomedical research as key enterprises of developed economies and – more importantly – improving health as a conditional good for each individual. That business organisations pursue their own economic aims does not contradict the value of the end "health promotion" as achieving collective ends can often only be accomplished by a different mode of implementation. Speaking generally, justifying an end and implementing it are different tiers of discourse you cannot separate, let alone identify! As biomedical research in general and the prospect of genomic research in particular promises to foster medical outcomes in quite a new way – it is arguable whether this might be assessed as an advance of a quantitative or qualitative dimension – each has to be accountable according to the classical test by the Golden Rule: do I or do I not want to derive benefit from what will admittedly not surely but probably become an advance? If in the case of a disease I want to take advantage of it, do I not violate the rules of reciprocity and mutuality if I intend merely to profit from it without being prepared to contribute towards practical implementation of the prospect? One may argue that we encounter in our lives many developments that we appreciate when they have become reality but which we have never supported in their implementation. That is true! Nevertheless there are two reasons why we should not assume an attitude of indifference to developments in genomics research, the success of which will probably depend on advances in biobanking. The first is the major importance of the objective of genomics research and, secondly, the great number of people necessary to accomplish it – imaginable breakthroughs in biobanking or pharmocogenomics, for example, will only be made with huge resources in the way of genetic and lifestyle data.

4 Some Notes on "Genetic Exceptionalism" and "Persons Genetically at Risk"

Before these visions can be subjected to ethical deliberation through adoption of a specific concept of obligation, some general fears have to be preconceived and climinated – not in general but as a general and as an undifferentiated concept: the concept of genetic exceptionalism (Murray 1997). It is necessary to discuss this briefly because of this: if it is true that genetic and genomic knowledge produces an exceptional and therefore allegedly an imminently dangerous knowledge, then it must be impossible to require moral obligations to participate in the field of genomics, and to reconcile this with the axiom of human dignity and communicative liberty. On the contrary, it would be of major importance to do everything to avoid moral obligations in this field.

But I am convinced that genetic exceptionalism is not only a misleading concept in general but one that produces inequities, especially in the area of postgenomics. Why? Obviously, lasting and – depending on the form – very precise predictions are special features of genetic knowledge. Moreover, its importance for reproductive decisions and the awareness of family traits results in a high symbolic and social explosiveness. Genetic knowledge is linked with fear of stigmatization and discrimination: short-term considerations of utility and benefit by insurances or employers; the memory of the inhuman eugenic practice in the first half of the 20th century which had a horrible climax in the Nazis' atrocities; finally, the cultural continuation of a crude genetic determinism and reductionism. Such fears give every reason to believe in the special character of genetic knowledge. However, from these arguments one cannot derive an exceptionality that might justify special treatment in dealing with genetic data, as compared with other medical procedures. Lasting and precise predictions, and the importance of genetic data for reproductive decisions and family knowledge, may to some extent characterise other biomarkers and medical or non-medical conditions as well. Only the density of these aspects and their cumulative effect permits us to characterize genetic and genomic knowledge as an area highly sensitive for the personal rights of individuals or groups in society.

Insight into the complex interaction between genome, internal biological processes and environment also contradicts the exceptionality approach. In dealing with diseases and sick people, how can genetic and other information be medically and legally separated in a clear way? Besides the difficulty of precise definition, one will be reproached for putting people without (explicit) genetic disorders at a disadvantage and giving preferential treatment to those who can prove a genetic illness. Such preferential treatment, in the form of increased protection, could result in a jealousy that confirms or reinforces social discrimination and stigmatization. Those who would differentiate genetic knowledge from medical information – even in order to hinder an advancing alleged "medicalization", "genetization" or "molecularization" of society – err in creating a genetic reductionism that should actually be avoided. Instead of treating genetic knowledge exceptionally, it would be better understood as one highly sensitive factor of medical knowledge. It should be seen as one brick in the desirable process of diagnosis and therapy on the individual level and on the level of Public Health. If other historic information is used for calculating risks in the field of life insurance or for discerning eligibility for employment in the occupational sector, then, to be fair, this applies to genetic knowledge as well, at least for the present.[7] The rule to be established is: let's talk about biomarkers, no longer about genetics. All depends on the predictive power of a biomedical diagnosis, regardless of the method by which it was obtained. Focusing only, or at least predominantly, on methods must be blamed for the inappropriate discrimination of methods.

---

[7] Though, what has to be refused and to be avoided is a forced genetic testing for the single purpose of taking out contracts. Health insurance coverage is a different case. It should provide everyone with a wide basic supply because health is a conditional good of life; cf. Schröder 2004, 285-317.

Before introducing the scheme for moral obligation to participate in genomics research, a new concept influencing the ELSA discussions on genetics and genomics must be taken into consideration. The concept of a "person genetically at risk" is worth addressing, as it may help to avoid some of the pitfalls of the criticized notion of genetic exceptionalism, and raises some different questions. In the following section I will argue that it should be applied – if at all – in a very restricted sense. I am convinced that this modification of genetic exceptionalism should by no means function as a counter-argument against the scheme of obligation in the field of genomics presented below. The notion of a "person genetically at risk" is allegedly characterized by the concurrence of three observed or constructed elements (Lemke 2008): the shift from "genetic destiny" to individual regulation of one's own body, secondly the shift from the prevention of conditions to the prevention of behaviours, and thirdly the effect of leaving the single individual alone in the land of nowhere between health and disease. Although I would not deny the emergence of new risks by genetic testing and genomic research I have to oppose for many reasons the idea of a person genetically at risk in the sense mentioned if it is proposed in a too general way.[8] A high risk, the first characteristic of the new concept, can only be alleged in the case of monogenetic diseases. The model cannot be applied to common complex disorders like diabetes, cancer or coronary heart disease, as it would not appropriately address the level of complexity resulting from the interlinkage of genome, environment and life-style. In the context of such diseases, genetic knowledge often has only a weak predictive effect in regard to the health status of person in question. Accordingly conclusions cannot be drawn in regard to evaluation of the societal situation or the development of protecting and enabling norms to the same extent that they may be drawn in the case of monogenetic diseases. On the contrary, in order to deal with diseases with multifactorial causes, genomic knowledge will be an integral part of a wide range of information and data. Genetic or genomic elements of common complex disorders cannot therefore in any practical sense be identified as a unique characteristic in understanding this complexity: each part has its own relevance. The picture will only be accurately drawn if one is mindful of all the very different colours.

The talk of the "person genetically at risk" must be seen against the backdrop of the fact that people will always have to make choices. It is the usual fact of human life that making one decision excludes other options.[9] We have had good cultural reason to move from the talk of fate and unproven authority to the scope of liberty and responsibility etc. Although I feel the risks of this movement I do not see a real alternative to a functionally differentiated and pluralist society that holds each person a bearer of human dignity. Human dignity will itself be preferentially

---

[8] For more detailed arguments cf. Dabrock 2008.

[9] According to Niklas Luhmann: "Draw a distinction!" is the basic imperative of knowledge and life at all; cf. Luhmann 2004.

proven by respect for autonomy in general, and autonomous decisions of individuals in particular, insofar as they do not infringe other persons.[10]

Therefore, one must seek alternatives to the concept of the "person genetically at risk". "Genetic fatalism" and "medical paternalism" will not be fruitful alternatives. I would agree with Rose and Novas who put forward the following statement years ago:

> „We shall argue that, far from generating resignation to fate or passivity in the face of biological destiny or bio-medical expertise, these new forms of subjectification are linked to the emergence of complex ethical technologies for the management of biological and social existence, located within a temporal field of 'life strategies', in which individuals seek to plan their present in the light of their beliefs about the future that their genetic endowment might hold. These new modes of subjectivity produce the obligation to calculate choices in a complex interpersonal field, not only in terms of individuals' relations to themselves, but also in terms of their relations to others, including not only actual and potential kin, past and present, but also genetic professionals and biomedical researchers" (Novas and Rose 2000: 488).

In their view this sober diagnosis is no reason for resignation but a shift of self-determination which must carefully be scrutinized. To manage these new challenges responsibly is not only an individual but also a societal task – no one should deny or underestimate this duty. It can however be confronted, and is no cause for alarm. The comprehensive idea of a person genetically at risk, when presented in a nonspecific way, endangers not only problematic, but also medically and socially fruitful, advances in biotechnology; it propagates an unwarranted hermeneutic of suspicion which cannot be attributed to the new pathways of genomics but which is still bound up in an old-fashioned understanding of genetics.

5 An Ethical Scheme for Obligations to Participate in Biobank Research

Having tilled the field by justifying the idea of an obligation to participate in some kinds of genomics research, and by dismissing the concepts 'genetic exceptionalism' and 'person genetically at risk' that hinder such an approach, we can now proceed to introduce the scheme of balancing an emerging conflict between respect for autonomy and common welfare in the field of genomics in general and in that of biobanks in particular. Put concretely, the question is 'what is the degree of obligation to take part in screening procedures, or to offer one's data to a biobank?' Cum grano salis, acknowledging the possibly colliding moral and legal goods of respect for autonomy and common welfare, I want to introduce the fol-

---

[10] This assumption does not exclude an idea of empowering persons for self-responsible decision-making via education.

lowing tentative analysis by an ethical step-by-step model.[11] It takes into account that these evaluative questions cannot usually be answered by yes or no.

A strong obligation arises on the part of the person concerned to take advantage of the reasonable results of a genomic vehicle, such as a test or a biobank endeavour, if it fulfils the criteria of efficiency or effectiveness including validity, reliability, specificity (Wilson and Junger 1968), so that great benefits can be obtained even with a limited amount of financial and intellectual input. Besides individual benefits such as avoidance of serious disease and promoting capacities in the individual for development of his or her own life plans, there is to also a big social benefit to be considered in terms of avoiding high material and immaterial costs that would result from delayed diagnoses or inadequate therapies caused by misdiagnosis. A further criterion, more relevant for assessment of the gradient of moral obligation, is an absence of any expectation of social stigmatization of the persons affected.

This alleged obligation may be set out in two ways. On the one hand, insofar as the criteria mentioned above apply, the society, usually represented by public health institutions, is obliged to provide and to guarantee these genetic and genomic means, not least in order to counteract the effect of the fact that they may be unavailable to the worst off. On the other hand, there is a strong moral obligation for the persons affected to participate in the respective procedures, especially regarding the comparably limited potential for harm to them and their families (little intrusion upon the formal right to self-determination, and a minimal tendency toward discrimination).

It is by no means of lesser ethical significance to put the rule the other way round, meaning to articulate it via negationis. Wherever the framing conditions mentioned above become weaker, the degree of moral obligation to take part in genetic or genomic heath care measures declines. For example counselling, established with good reason as a procedure to accompany individual data-gathering in genomics, should be more non-directive. To put it differently, to refuse participation requires less moral reasoning.

Another distinction relevant for discerning when to put the idea of obligation into practice is that although participation in genomics research and testing can be read as a moral imperative under the conditions referred to, it does not have to be immediately transformed into legal compulsion. The proven sense of a legal culture based on negative liberty and informed consent raises the idea of keeping the principle of voluntariness on a legal level without restricting oneself to a standardized non-directive counselling. Simple ethical deductions do not work here. Different social points of view must be recognized, because moral questions in a moral-political discourse cannot be answered deductively. They must rather be balanced with cultural standards, which is the only way to secure the acceptance and reproducibility of moral decisions, which are as important as moral validity.

---

[11] Cf. Brand et al 2003.

Incidentally, a question comes up that refers to health economics and applied justice: is it actually possible to maintain the high standard of non-directiveness in counselling, as it is known in medical human genetics? This is probably hardly possible. Then establishment of the extent of counselling under the very different, not so problematic, conditions of common complex disorders, as compared to counselling in the case of other, especially monogenetic, diseases, means not only to take into account economic forces, but also and even more so the depth of intrusion into the informational self-determination.

This suggested rule of priorisation does not constitute paternalism in health care that would be unacceptable for a pluralistic society. The contrary is true: only by guaranteeing elementary conditions (so-called conditional or primary goods) can different ways of successful life be realized. A consideration of liberty without acknowledgment of these primary goods will get us nowhere. It would even be capable of use against the disadvantaged who are unable to arrange their liberty because of what is in reality a mere formal ideology of liberty that does not provide equality of opportunity for them.

The consideration of social duties such as contributions to corresponding projects may be relevant in relation to the pursuit of justice on behalf of the underprivileged. Without imposing a strong moralizing undertone, those of lesser means can help to make people more sensitive to the fact that their lifestyles rest on presumptions of which they are completely unaware. In light of economic discrepancies, it really makes a difference whether someone may personally benefit from a genetic analysis for a test or a biobank or whether a close relative or society in general is concerned. It also makes a difference whether there are relatively low or high chances of success. Chances are, however, often assessed inadequately, since the highest potential sometimes lies in those projects that seem very complicated at first glance. Strong determination is required to achieve a breakthrough by apparently complicated methods. All these aspects must be taken into account in a discussion about potential moral obligations. It certainly does not make sense to assign more moral relevance to (seemingly) easily achievable aims or to urgent matters. The objection to this argument formulates a false alternative. It is criticized that moral obligations are fixed for comparatively unpromising medical research, while easily realizable possibilities for improving the state of health of many people and particularly of the underprivileged are ignored. This alternative can, of course, be promoted by fighting for a more just decent minimum guaranteed by the welfare state – finally, you will see that progress in genome-based medicine is indispensable for further improvement in the wide field of endemic and infectious diseases such as malaria and AIDS. There will be no such enhancement if corresponding data are not made available, i.e. if there are no biobanks storing and processing the data of great numbers of participants. Our abovementioned rule also holds here: do A and still do B. This leads me to the overall aim of my discussion, which is to show how fruitful it is to formulate the reasons and clear-cut limits of moral obligations for participation in genome-based research. The observation that there are, in fact, tangible criteria for justifying participation or refusal of participation, may strengthen trust in such projects. The common suspicion that a dark, impersonal "biopower" might take possession of innocent, exposed and defenceless individuals does not need to be treated here in detail: there are good reasons for defining limits of obligations, taken that you base these definitions on a differentiated concept of 'obligation'.

Nevertheless, and this will be my last short point: the perpetuation of fears, though they are not as warranted as in the past, given that the genomic hype has come down to earth, dramatically illustrates the need for building of a greater degree of trust in the field of genomics. Building up trust would seem to be the major obligation for all those institutions that are interested in fostering individual and public health outputs, not only effectively but also responsibly. Systematically speaking, building up trust is not only a case of smoothly applying ethical principles but an intrinsic moral dimension in itself.

Maybe on this point we could reach agreement between those who accept the concept of the person at genetic risk and those who are more interested in fostering a responsible shaping of medicine to facilitate the integration of genomic knowledge within it. Concerning the modes and levels of building of trust – I suppose – we still have reason for forthcoming dispute.

References

Brand A, Dabrock P, Gibis B (2003) Neugeborenen-Screening auf angeborene Stoffwechselstörungen und Endokrinopathien – aktuelle ethische Fragen aus unterschiedlichen Perspektiven In: Dörries A et al (eds) Das Kind als Patient – Ethische Konflikte zwischen Kindeswohl und Kindeswille. Campus, Frankfurt am Main, 217-233

Dabrock P (2008) Risikodimensionen genetischer Tests bei Adipositas. In: Hilbert A, Dabrock P, Rief W (eds) Gewichtige Gene. Adipositas zwischen Prädisposition und Eigenverantwortung. Huber, Bern, 167-190

Dabrock P, Klinnert L, Schardien, S (2004) Menschenwürde und Lebensschutz. Herausforderungen theologischer Bioethik. Gütersloher Verlagshaus, Gütersloh

Dagger R (2007) Political Obligation. The Stanford Encyclopedia of Philosophy (Summer 2007 Edition). http://plato.stanford.edu/archives/sum2007/entries/political-obligation

Heyd D (2006) Supererogation. The Stanford Encyclopedia of Philosophy Fall (2006 Edition). http://plato.stanford.edu/archives/fall2006/entries/supererogation

Johnson R (2008) Kant's Moral Philosophy. The Stanford Encyclopedia of Philosophy (Summer 2008 Edition). http://plato.stanford.edu/archives/sum2008/entries/kant-moral

Kant I (1959) Foundations of the Metaphysics of Morals. Bobbs-Merrill Publishing, Indianapolis

Lemke T (2008) Von der sozialtechnokratischen zur selbstregulatorischen Prävention: Die Geburt der „genetischen Risikoperson". In: Hilbert A, Dabrock P, Rief W (eds) Gewichtige Gene. Adipositas zwischen Prädisposition und Eigenverantwortung. Huber, Bern, 151-165

Luhmann N (2004) Einführung in die Systemtheorie. Carl Auer, Heidelberg

Murray T (1997) Genetic Exceptionalism and 'Future Diaries': Is Genetic Information Different from Other Medical Information? In: Rothstein M (ed) Genetic Secrets: Protecting Privacy and Confidentiality in the Genetic Era. Yale University Press, New Haven, 60-73

Novas C, Rose N (2000) Genetic Risk and the Birth of the Somatic Individual. Econ Soc 29: 485-513

Nussbaum M (2000) Women and Human Development. The Capabilities Approach. Cambridge University Press, Cambridge

Nussbaum M (2006) Frontiers of Justice. Disability, Nationality, Species Membership. Harvard University Press, Cambridge et al

Powers M, Faden R (2006) Social Justice. The Moral Foundations of Public Health and Health Policy. Oxford University Press, Oxford

Rabinow P (1996) Artificiality and Enlightenment. From Sociobiology to Biosociality. In: Essays on the Anthropology of Reason. Princeton University Press, Princeton, 91-111

Rose N (2007) The Politics of Life itself. Biomedicine, Power, and Subjectivity in the Twenty-First Century. Princeton University Press, Princeton

Schröder P (2004) Gendiagnostische Gerechtigkeit. Eine ethische Studie über die Herausforderungen postnataler genetischer Prädiktion. LIT Verlag, Münster et al

Wilson JMG, Junger G (1968) The principles and practice of screening for disease. World Health Organisation, Geneva

World Health Organisation (WHO) (2002) Genomics and World Health. Report of the Advisory Committee on Health Research. World Health Organisation, Geneva

Donors and Users of Human Tissue for Research Purposes

Conflict of Interests and Balancing of Interests

Christian Lenk

Abstract Research with human tissue has often no direct benefit for the participating patients or test persons. Researchers, who want to use human tissue for their projects have to seek mechanisms to motivate tissue donors which are practicable in the framework of a specific study design. As some former cases show, there is a potential conflict of interests between researchers and patients which sometimes even culminates in legal proceedings with counterintuitive outcomes. The article presents two prominent cases and analyzes notions of justice for the field of tissue extraction and research and the distribution of possible benefits to researchers and patients. The Aristotelian conception of justice and distribution serves as a starting point for this analysis. Criticism concerning the commercialisation of the human body is taken into account and a variety of commercial as well as non-commercial forms of benefit-sharing are proposed and discussed with regard to the specific character of the research study. Thinking the categories of study type and potential benefits together results in a chart with customized benefits for each research study. The article comes to the conclusion that forms of benefit-sharing are an adequate instrument to avoid conflicts of interests in projects which are based on research with human tissue. It seems to be important to choose the right form of benefit-sharing for a specific project, dependent on the project's characteristics and potential benefits for the patients.

1 Introduction

Research with human tissue and body material in the framework of biobanks needs ethical and legal regulations as unambiguous as possible to cover the need for tissue samples and to protect the interests of tissue donors and researchers. In contrast to the relationship of patient and physician, who generally have the common objective of healing the patient's disease, the relationship of ill or healthy tissue donors and researchers or research institutions is different. This is due to possible conflict or incongruency of interests in the case of the extraction and scientific use of human tissue and body material. These conflicts are different in different constellations of research, according to the different frameworks of academic or commercial research and research which aims at an individual or a collective benefit for patients. Prima facie, one can distinguish the following constellations:

- a patient is asked in the course of his therapy to donate tissue for research purposes which aim at the development of a new treatment strategy for the patient's disease;
- a patient is asked in the course of his therapy to donate tissue for research purposes which aim at the development of a new treatment strategy for another disease or for basic research;
- a healthy proband is asked to donate tissue for basic research and to connect his genetic data with personal health information;
- patients or probands are asked to donate tissue for the commercial development of new pharmaceuticals.

The concerned interests are of a different type, i.e. on the side of the patients the vital interest for the development of an effective treatment of their disease, on the side of the researcher the smooth carrying out of high-quality research and on the side of commercial research the development of new drugs in the framework of a given business model. Therefore, it must be the task of medical ethics to ask how the described interests can be balanced in a way that is fair according to theoretical concepts of justice. However, there exists the problem that the most influential current piece of theory, John Rawls' Theory of Justice, deals more with life opportunities of individual persons and justice in societal institutions, and less with concrete questions of distributive and corrective justice. The locus classicus for the solution of such rather concrete questions are the traditional approaches to justice, i.e. iustitia distributiva and iustitia commutativa (distributive or corrective justice, cp. Aristotle, Nic Eth., V5, 1130b) in the work of Aristotle.

The classic and normatively superior model in the area of organ transplantation is the so-called donation model, i.e. the living or dead donor gives an organ for the enhancement of another person's health condition without financial compensation. However, there seems to be a growing uneasiness in the medical ethics community whether this model, which stems from a context of therapy should be transferred without modifications to the context of basic research or commercial research and product development. For example, as Eve-Marie Engels argued in her introduction to the annual conference of the German National Ethics Council (Nationaler Ethikrat):

> The principle of the non-commercialization of the human body is a legally anchored norm or unwritten law in many countries. However, there is need of clarification on what this principle should be applied. When it is relevant for all body substances and connected data and information, it can happen that all participants (researchers, industry, etc.) profit economically from this research which bases on the donation of samples, ex-

cept the altruistic donors themselves, may it be individual donors, groups or populations[1] (Engels 2002: 20f).

Indeed, this critical perception is supported by two cases of conflict of interests which will be discussed in the next paragraph. This is the reason of the development of diverse alternative models for joint cooperation of researchers and patients with the goal of an adequate balancing of interests. It is the objective of this article to evaluate how this goal of balancing interests can be achieved within the methodology of medical ethics.

2 Examples from the Practice of Biomedical Research

In this paragraph, two famous legal cases will be discussed, i.e.

* Moore v. the Regents of the University of California and
* Greenberg v. Miami Children's Hospital Research Institute.

However, interesting from the point of view of medical ethics is not the legal evaluation of these cases in a specific national dispensation of justice, but rather the kind of ethical problems which emerge in until now rather unfamiliar fields of medical research. An interesting philosophical side-effect is the question of the commodification of human body parts which has some interesting connections to the philosophical discussion on civil rights and property in the era of the enlightenment, i.e. how can anybody possess tissue or private information from another person and which rights regarding these things remain at the original person (Lenk 2008).

2.1 The Moore Case

John Moore was a leukemia patient in the mid of the 1970s and was treated at the medical facilities of the University of California (UCLA). After the curing of his initial disease, he was asked by his physicians David Golde and Shirley Quan, to visit the hospital again and again (from 1976 until 1983) for further examinations and the taking of blood serum, bone marrow and sperm samples. The reason for

---

[1] „In vielen Ländern gilt das Prinzip der Nichtkommerzialisierbarkeit des menschlichen Körpers, sei es als gesetzlich verankerte Norm oder als ungeschriebenes Gesetz. Klärungsbedürftig ist jedoch, worauf sich dieses Prinzip beziehen soll. Wenn es sich auf sämtliche Körpersubstanzen und die damit verknüpften Daten und Informationen erstreckt, so kann es geschehen, dass alle Beteiligten (Forschende, Industrie usw.) wirtschaftlich von dieser auf der Spende von Proben basierenden Forschung profitieren, außer den altruistischen Spendern selbst, seien es individuelle Spender oder Gruppen und Populationen." (translation by the author).

this further visits was not a therapeutic one, but the physician's discovery that Moore's spleen produced a protein which had some potential for cancer therapy. Therefore, the physicians produced, without Moore's knowledge, a stem cell line from his bodily material and patented it as their own invention. For the commercial exploitation of the cell line, the physicians sold licenses to Sandoz and Genetics Institute Inc.. After Moore learned by chance that a cell line was produced from his samples, he sued the UCLA for a monetary compensation (cp. Hoppe 2007, 200 ff.)

The case produces some interesting questions: was the bodily material originally John Moore's property in the usual sense of tangible property? Were the tissue and cells later on the property of the physicians and the UCLA? What was the decisive contribution to the creation of the cell line – the raw material which was extracted from the patient or the invested research work and medical knowledge? Did Moore experience any harm and were all his personal rights respected? Which party should have the benefit from the financial exploitation of the cell line? Surprisingly, the patient John Moore was not successful before the California Supreme Court to get compensation for the breach of confidence of his deceitful physicians and had to reach a settlement with his opponents. Intuitively, this seems to be unsatisfying from the point of view of corrective justice.

2.2 The Greenberg Case

In the year 1987, Daniel Greenberg, a father of two children with Canavan disease searched a scientist who could find the genetic cause of his children's disease. Canavan disease is a hereditary, autosomal-recessive degeneration of the central nervous system which manifests in early childhood. The disease occurs statistically in 1 of 6400 Ashkenazi Jewish children. The Greenberg case can therefore count as an example for research on a disease which affects a special group of patients. Greenberg contacted the physician and scientist Reuben Matalon who agreed to work together with other families which were affected by Canavan disease. The concerned families themselves supported the researcher with samples from their children, a large amount of money and the establishment of the first registry for patients with Canavan disease. Matalon was successful and found the relevant gene in the year 1993. This was the precondition for the development of a genetic test for Canavan disease which enabled the Canavan Foundation to give potentially concerned persons free access to genetic testing in the year 1996. In the following, Reuben Matalon and his employer, the Miami Children's Hospital, patented the gene and the genetic test without the knowledge of the families or the Canavan Foundation. Therefore, the genetic test was know commercialized and everybody who wanted to use it had to apply for a license at the patent holders (Marshall 2000, 1062). As the Canavan Foundation commented:

The gene which had been found through the support of the Greenbergs and the other Canavan families, the gene which came from the tissue of their dead children, the gene which the Greenbergs and other plaintiffs had worked to find in order to save other families from their heartache, had now been patented by Dr. Reuben Matalon and Miami Children's Hospital (Canavan Foundation, 2000).

Unfortunately, Daniel Greenberg lost the legal proceedings equally like John Moore because the Florida District Court decided that it was the researcher's and the hospital's right to patent the gene and the genetic test.[2] Obviously, the main point of dispute in this case is not the compensation for a financial damage, but rather the right to control and to secure the access to the results which were produced by the cooperation of the patients and the researcher. Taking into account that the whole research project was originally initialized by the concerned families, the court decision leads to a real expropriation of the patient's families.

As a summary, it can be noticed that substantial doubts whether the court decisions and the dealings of the physician-researchers with their patients were appropriate according to the standards of medical ethics in general and the standards of justice in particular remain. As Nils Hoppe remarked[3], and like it is also stated in the citation from the Canavan Foundation (which mentions the "gene", i.e. the genetic information), the decisive point seems not to be who the legal owner of the tangible property of the cells and samples is, but rather who has legal or moral claims for the intangible or intellectual property which is derived from the cells and which is finally patented. This leads to a possible analogy from the area of intellectual property rights and to the question, as it will be shown in the following, whether it was not simply the superiority of the stronger party (which are usually not the patients or the consumers) which lead to the legal decisions.

Therefore, I want to present the following two fictitious cases with the question whether they represent an analogy or not:

> Person A gives person B an object for the purpose of non-commercial research. B does not follow the defined purpose and makes a financial benefit. Does B have to give person A a part of the money or the whole amount of money? Has B done something morally or legally wrong?

> The enterprise A sells intellectual property to person B for private purposes. B does not follow the defined purpose and makes a financial benefit by selling the intellectual property to further persons. Does B have to give the enterprise A a part of the money or the whole amount of money? Has B done something morally or legally wrong?

---

[2] United States District Court, S.D. Florida, Miami Division, Court Decision: 264 Federal Supplement, 2d Series 1064; 29 May 2003 (date of decision).

[3] Hoppe 2007, 203: „Golde and Quan put some considerable, artful work into preparing a cell line wich – by virtue of the work invested – suddenly becomes capable of ownership. Whilst this logic clearly holds true for a number of reasons, it disregards one very important aspect of *Moore*: the real value of the claim lies in the intangible property that was extracted from the cells."

The two cases are taken from the area of medicine and the field of intellectual property. If one sees an analogy here, it has to be demanded that both cases should be evaluated in a similar way. This would mean, that patients or probands, who give samples for a specified purpose should be compensated when their samples are used for other, especially commercial, purposes. The question is, why the "change of purpose" without the information of the original owner could be acceptable in the first case, but not in the second. In both cases the decisive point is the expected probability of the new owner keeping the joint agreement not to commit to commercial activities. One could argue, that in the second case the seller has to prevent the resale of its intellectual property because this would be a financial damage for him. However, if the donor in the first case would have known about the commercial possibilities, it would have been also a potential financial damage for him. It is therefore useful to get deeper into the topic of distributive and compensatory justice to see what the application of this normative framework would mean for the ethical problems described.

3 Justice and Distribution of Goods in the Aristotelian Tradition

Aristotle describes injustice in the Fifth Book of the Nikomachean Ethics as "the breaking of laws or wanting to have more (than other persons), and thereby disrespecting equality."[4] In the following, he describes two types of justice, i.e. distributive justice and compensatory justice which can also be understood as a "justice of exchange", what makes this type of justice relevant for the question of the present article. The notion of equality is central for Aristotle's thinking about justice, which seems a little confusing from the modern point of view, regarding the fact that according to this ethical framework, one can only be fair towards other male, full citizens of a state or polis, but not, for example, towards women, children and slaves. Therefore, one only has to act fair towards people who have the same, consolidated status as oneselve. It is highly evident, that the enlightenment produced a paradigm change in that regard and we see it nowadays as a kind of discrimination to exclude certain persons from fair treatment.

The described examples from biomedical research in the last paragraph let it seem reasonable to suppose that the iustitia commutativa is the most relevant type of justice for the present normative problem. But this is only true if one thinks about the problem and the justice of exchange ex post, i.e. when the problem of injustice has already occured. But the iustitia distributiva seems to be more fundamental for the question, what the appropriate part which each party should receive for his activities is. This perspective would fit better to the evaluation of cases or the solution of problems of justice which are still in the future. When this presup-

---

[4] Cp. Nikomachean Ethics V, 2, 1129a32-34; Gordon 2007, 39 f., Fn. 32 (translation by the author).

position is true, another problem occurs: the approach of iustitia distributiva can only be applied, if there is a kind of *ton koinon*, i.e. a common good, which has to be distributed, like the common good of the state which is distributed to the citizens according to their needs and contributions. Aristotle demands for the iustitia distributiva that the citizens of the state should receive rewards from the common good in proportion to the contributions which they have given to the state (Gordon 2007: 156).

Therefore, we have to ask for the solution of our normative problem, whether there is a common good or common activity between the researcher and the patient which establishes a similar relationship between these parties. And obviously, there is indeed a common activity and a common good, at least in the two cases that were described in the last paragraph. In the two cases, the researchers treated their medical project as a kind of financial investment and privatized the financial earnings by patents on the cell line, the gene or the genetic test. The privatization functioned as a kind of exclusion of the patients from the project. However, there can be no doubt, that before this exclusion, the researchers were dependent on the contributions of the patients for the success of their project.

Beneath the mathematical considerations, how one can define the right proportion of the distribution and allocation of goods, there is a very explicit example which Aristotle gives in the Nicomachean Ethics for the creation of distributive justice:

> When a sum of money shall be distributed and a number of persons have a claim to it, this should be done in the same ratio like the contributions which the individual persons delivered. Injustice, which is the opposite of justice in this sense, is what violates this proportion.[5] (Aristotle 2004: 102).

Aristotle presents this as an example for compensatory justice in business connections. Fundamental for this principle of compensation is the precondition that the individual partners, in principle, have equal rights and their resulting benefit should be in relation to their contribution or investment to the project. The just proportion would be, that a person which invests 10% of the total expenditure for an enterprise should also receive 10% of the earned benefits. Although the example mentions commercial relationships, this does not necessarily have to be the case. There could well be non-commercial commitments or investments, which would not change the basic idea of the "just" proportion.

Therefore, we can conclude, that the patients should receive a reward from this joint project in proportion to the contribution which they have made. When two parties participate in a common project, and the contribution of the first party equals 15% and the contribution of the second party amounts to 85%, the profit

---

[5] „Wenn eine Geldsumme, an die mehrere einen Anspruch haben, verteilt werden soll, so wird es nach demselben Verhältnis geschehen, in welchem die Beiträge zueinander stehen die jeder geliefert hat; und das Ungerechte, das den Gegensatz zum Gerechten in dieser Bedeutung bildet, ist das was gegen die Proportion verstößt." (translation by the author).

from the project should be distributed accordingly. In the Moore case, this could have been an amount of money, to compensate the plaintiff for his troubles in the compulsive participation in the research project, or free access to the genetic test in the Greenberg case, to give an adequate reward to the concerned parents for the invested money and their activities. It may be easiest to calculate the adequate reward for the invested money in the case of a project's financial success, but it is surely no problem to transform the kind of participation of patients in commercial medical studies into an adequate reward, according to the findings and possibilities of a specific project.

But this solution would not take the critique concerning the commercialisation of research and of financial incentives towards patients into consideration. As mentioned above, this critique seems to stem from the area of organ transplantation and it is not clear, whether it should be transferred to the area of research with human tissue. Additionally, the critique of the commercialisation of research remains one-sided following the quotation of Eve-Marie Engels cited above. Why are only the patients concerned about this critique and not the researchers themselves? However, one important argument remains against financial incentives for patients in research. Patients should commit to research and make their decision not because of financial incentives, but because they think that it is reasonable and valuable for their kind of disease. This does not mean that they should not be compensated financially ex post, in the case of a violation of their personal rights – but it means that they should get an offer for an individual financial share of the project's profit.

4 Options for Financial and Non-financial Benefit-sharing

Therefore, it is useful to think which alternatives exist for a possible balancing of interests between patients and researchers. From my point of view, there are the following possibilities which could be ruled by a contract between the research institution or enterprise and individual patients or groups of patients:

- *Access to innovative therapeutic or diagnostic methods:* It is a principle of research ethics, which is also expressed in §30 of the Declaration of Helsinki, that every patient who participates in a research study should finally have access to the newly developed therapeutic and diagnostic measures.[6] This possibility should, from my point of view, have been directly applied to come to a balance of interests in the Greenberg case.
- *Individual feedback to participants:* Research with biobanks is a special case of medical research. Two kinds of projects can be distinguished, i.e. research pro-

---

[6] World Medical Association, Declaration of Helsinki, § 30: "At the conclusion of the study, every patient entered into the study should be assured of access to the best proven prophylactic, diagnostic and therapeutic methods identified by the study."

jects which aim at one or more specific diseases and population genetic bio-banks. In the first case, patients will have to give samples to the biobank which will store these samples and coordinate the research activities. Therefore, it would be possible to define specific genetic findings which are relevant for the individual medical therapy, which should be communicated to the patients. For example, when in the course of a genetic screening tumor markers are found which show that the patient has cancer, he should be informed that it is possible to start a therapy. From my point of view, it is necessary to define these possible information before the project starts and to give only those information to the patients which are of high therapeutic relevance, but not mere genetic probabilities. In the second case of population genetic biobanks it seems less probable that a concrete advice for the disease of an individual patient is found. However, maybe there are research projects which work with the biobank's material which are quite similar to the projects described in the first case. In those cases, there should equally be a duty to inform the patient about findings which are highly relevant from the point of view of therapy. The duty to inform patients about concrete findings which are relevant for therapy should be taken very seriously, because in the case of a violation of that duty research projects or biobanks could be sued by patients because they withheld the information and damaged the patient's health because he was not able to start an adequate therapy. Therefore, it has to be seen critical that, for example, the U.K. Biobank rules out individual feedback to participants in its Ethics and Governance Framework.[7]

- *Investments in medical infrastructure:* There are regularly discussions in the area of medical ethics concerning research in developing countries or research with poor patients. The gathering of samples in developing countries for genetic analysis can be seen as one area of such research. Examples are research projects with the Hagahai people from Papua New Guinea or the Guyayami in Panama (Widdows 2007: 168 f.) Therefore, the Statement on Benefit-Sharing of the HUGO Ethics Committee recommends concrete investments into the communal and medical infrastructure in such areas:

> Moreover, immediate benefits such as medical care, technology transfer, or contribution to the local community infrastructure (e.g. schools, libraries, sports, clean water etc.) could be provided. In the case of profit-making endeavours, the general distribution of benefits should be the donation of a percentage of the net profits (after taxes) to the health care infrastructure or for vaccines, tests, drugs, and treatments, or, to local, national and international humanitarian efforts (HUGO Ethics Committee 2000).

Those investments do not seem to make much sense in the case of the classical, continental European welfare states like for example the Scandinavian countries where every citizen has access to appropriate health care. But it definitely makes sense in developing countries where this is not the case and it could even be taken

---

[7] U.K. Biobank 2003, see paragraph (I.B.3.c); cp. also Brownsword 2007 on this issue.

into consideration in post-industrial states with strong neoliberal policies where large parts of the population do not have adequate access to appropriate health care. In the case of local communities in those states which have to suffer from rudimentary health care, the same argument would apply.[8]

- *Improvement of health care for a specific patient group:* In the case of chronic or rare diseases, it may be an option to share research benefits and balance the interests of patients and researchers by an improved health care for a group of patients. This possibility will normally be limited to patients in one region, but can act as a model in the case of a successful cooperation between patients and researchers or in the case of strong benefits for the concerned patients. One example for the improvement of health care for a specific patient group is the case of asthma research with the inhabitants of the small island Tristan da Cunha which is located in the South Atlantic. A Canadian scientist found that nearly 50% of the island's inhabitants suffer from asthma. Because the inhabitants were descendants from British sailors and settlers, and in the very small group of less than 300 persons exist only seven or eight different families, the probability was high to identify the genetic background of asthma with the help of the inhabitants. Samples and health records of 289 inhabitants were taken in 1993. Later on, the genetic material was analyzed by the company Axys Pharmaceuticals which patented a number of genes. The search for the 'asthma gene' was financed by the German company Boehringer Ingelheim, which produced a knock-out mouse with the obtained genetic information. The inhabitants from Tristan da Cunha received an expense allowance for their participation, but additionally the local hospital was modernized on the expense of Boehringer Ingelheim. It is not known, whether the inhabitants of the isle find this agreement fair from the present point of view, but it can serve as an example for a collective benefit for a specific patient group by the improvement of health care (Scott 2003; Slutsky and Zamel 1997).
- *Collective sharing of financial profits*: In the case of commercial research projects which need the cooperation of a group of people and where no other kind of benefits are senseful, the sharing of financial profits should not be categorically excluded. As was outlined above, the normative framework of iustitia distributiva demands the proportional sharing of benefits from joint projects. The decisive point seems to be, that there is no exclusive incentive (or external

---

[8] In the discussion during the conference at the University of Marburg, which was the starting point for the present volume, Agomoni Ganguli raised the (important) question whether this specific form of benefit-sharing can be justified with the Aristotelian approach of proportional rewards. The question could be, whether this kind of benefit-sharing is not rather advisable due to other arguments, like for example that everybody should have a "decent minimum" of health care. However, the "decent minimum"-arguments points from my point of view stronger to the duty of the state, not to the duty of a partner of a joint project. The specific obligation of the coordinator of a research project would be indeed an adequate return for the participation at the project – and from the point of view of marginized groups it would be highly recommendable if this would result in an adequate health care.

pressure) for individual patients to participate in a study and to give samples. But this is not necessarily the case when, for example, a specified group of patients or people from a special community decide to participate collectively in a research project and receive a financial reward for this activity. The amount of money should be calculated in relation to the patients' or participants' expenditure and the project's financial profit.

5 Implications for Different Constellations of Research with Biobanks

The article started with a short description of four different constellations which are normally distinguished in the normative evaluation of medical research: research with a possible individual benefit, research with a benefit for other patients, basic research with healthy subjects and finally commercial research. Usually, the decisive question in the ethical evaluation of such research scenarios is the assessment of the connected risks and benefits for patients. One can say, that the expected bodily risks for the donation of samples of body material are low. However, this is not the only kind of risk, given the special nature and relationship of genetic material to the donor. For example, personal genetic information which can be analyzed by employers or private insurances are not in the best interest of patients and subjects. Although most biobanks will limit access to the material of scientific projects, it can never be fully excluded that there might be access to genetic information for persons with other interests. It could also be the case that the police or other state authorities try to get access to biobanks to identify persons who left genetic material at a scene of crime. Therefore, it remains a small, but important risk for sample donors in such projects which lies more in the context of personal rights, informational self-determination and a possible discrimination as a carrier of a specific gene than in the context of definite bodily harm.

On the other hand, the benefits for patients or subjects in biobank research are rather limited or depend on the nature or guidelines of specific research projects. Here one can see an analogy to §30 of the Declaration of Helsinki which demands that all patients should receive the best possible treatment after the conduction of the study. This means, that for example patients who were allocated to a study's placebo arm receive, ex post, a benefit from the study for their participation. Equally, it can be demanded for medical studies with biobanks, that patients should receive a benefit for their participation and that this benefit should result from their personal health condition and the character of the project. If there are possible benefits for the individual health of patients, like for example information about tumor markers, these information should in all kinds of projects be given to the patient. In the case of group-specific or commercial research projects, the benefit should equally be group-specific or should include financial support.

Therefore, one can put together possible benefits for different kinds of studies on the following table:

Table 1: Possible benefits from biobank research and different study types

Character of the study	Access to innovation	Individual feedback	Investment in infrastructure	Improvement specif. group	Financial profits
Study with poor patients	X	X	X		
Group-spec. study	X	X		X	
Academic research	X	X			
Commercial research	X	X			X

Critics may point out, that this would result in too many administrative or financial burdens on the shoulders of researchers and research institutes. However, one has to see that in many research projects the researchers need the active cooperation and a lot of personal data from their subjects of research and the experience of individual benefits on the side of the patients will create additional trust in the seriousness of research projects. Additionally, the expected benefits always have to be seen in the framework of the administrative and financial possibilities of specific projects. But it is necessary, from my point of view, that the demands of the Declaration of Helsinki are not undercut and that a possibility for individual feedback of highly relevant health information exists. Although there may be projects in which neither the one nor the other benefit is realizable, there are surely a large number of other projects where it is possible and, according to the present arguments, benefits should be implemented in the study design.

6 Conclusions

The present article dealt with a number of cases of research with biobanks in which a conflict of interests between researchers and patients exists. A special problem seems to be the sometimes clandestine commercialization and privatization of research projects and the patenting of findings which restricts the access to benefits from research projects for participating patient groups. But there are also a number of mechanisms for balancing interests that can be combined to a binding ethical and legal framework for benefit-sharing in the course of such projects. The main points on the way to such a framework are, from my point of view:

- Patients or probands should be fully informed about the academic or commercial character of a research project and the purpose for which a sample was taken.
- If the information was wrong or misleading, the samples should be destroyed and the patient or proband has to be compensated for the malicious deceit and the violation of his personality rights.
- The character of a research project and the situation of the patients or probands are decisive for the kind of benefit which is appropriate for the participants.

- Access to innovative diagnostic, preventive or therapeutic methods or an individual feedback on a patient's health condition are recommendable for all kinds of research projects with biobanks, insofar as such methods are developed or such information are generated in the course of such a project.
- Although individual financial incentives for study participation are potentially dangerous in research projects, there are a number of effective methods for collective benefit-sharing which make sense and are reasonable from an ethical point of view.

Acknowledgments I have to thank Nils Hoppe from the Working Group for Medical Law and Bioethics of the Leibniz University Hanover for discussions and valuable insights into the topic in the course of the joint preparation of the Tiss.EU project which will be funded by the European Commission in the 7th Research Framework Programme. He also gave me the information about the asthma project on the island Tristan da Cunha.

References

Aristotle (2004) Nikomachische Ethik. Cited after the edition of the Digitale Bibliothek in the translation of Adolf Lasson, Berlin

Brownsword R (2007) Biobank Governance: Property, Privacy and Consent. In: Lenk C, Hoppe N, Andorno R (eds) Ethics and Law of Intellectual Property. Current Problems in Politics, Science and Technology, pp 11-25. Ashgate, Aldershot

Canavan Foundation (2000) Canavan Foundation Joins Lawsuit against Miami Children's Hospital. http://canavanfoundation.org/news/10-00_miamihostpital.php. Accessed 06 September 2007

Engels EM (2002) Biobanken für die medizinische Forschung – zur Einführung. In: Nationaler Ethikrat (ed) Biobanken. Chancen für den medizinischen Fortschritt oder Ausverkauf der ‚Ressource' Mensch? pp 11-22. Nationaler Ethikrat, Berlin

Gordon JS (2007) Aristoteles über Gerechtigkeit. Das V. Buch der Nikomachischen Ethik. Alber, Freiburg, München

Hoppe N (2007) Out of Touch: From Corporeal to Incorporeal or Moore Revisited. In: Lenk C, Hoppe N, Andorno R (eds) Ethics and Law of Intellectual Property. Current Problems in Politics, Science and Technology pp 199-210. Ashgate, Aldershot

HUGO Ethics Committee (2000) Statement on Benefit-Sharing. www.hugo-international.org/Statement_on_Benefit_Sharing.htm. Accessed 19 September 2007

Lenk C, Hoppe N (2007) Ein Modell zur Konstitution von Nutzungsrechten an menschlichem Gewebe und Köpermaterialien. In: Taupitz J (ed) Kommerzialisierung des menschlichen Körpers pp 199-211. Springer, Berlin, Heidelberg

Lenk C 2008: Gibt es das Recht auf Eigentum am eigenen Körper? Ein Beitrag zur Forschungsethik in der kantischen Tradition der Aufklärung. Z Med Ethik 54: 13-22

Marshall E (2000) Families Sue Hospital. Scientist for Control of Canavan Gene. Science 290: 1062

Rawls J (1971) A Theory of Justice. Belknap Press, Cambridge

Scott S (2003) ‚Loneliest island' may hold key to asthma. National Post, 18 January 2003

Slutsky AS, Zamel N (1997) Genetics of Asthma. The University of Toronto Program. Am J Respir Crit Care Med 156: 130-132

U.K. Biobank (2003) Ethics and Governance Framework. Background Document. www.wellcome.ac.uk/assets/WTD003287.doc. Accessed on 18 September 2007

Widdows H (2007) Reconceptualizing Genetics: Challenges to Traditional Medical Ethics. In: Lenk C, Hoppe N, Andorno R (eds) Ethics and Law of Intellectual Property. Current Problems in Politics, Science and Technology pp 159-173. Ashgate, Aldershot

Collection of Biospecimen Resources for Cancer Research

Ethical Framework and Acceptance from the Patients' Point of View

Johannes Huber, Esther Herpel, Frank Autschbach, Stephan Buse, Markus Hohenfellner

Abstract The collection of tissue samples has become a valuable basis for medical research. Relying on a deeply rooted tradition in cancer research, a common biobank project has been established in Heidelberg within the "National Center for Tumor Diseases" (NCT). As there is a lack of specific legal regulation in Germany, organisational requirements are uncertain. The use of a one-time general informed consent procedure containing both oral and written elements, attended by a physician, should guarantee the utmost legal security and respect for ethical considerations. The project is also mentioned in the hospital treatment contract. As the idea of 'informed consent' has become indispensable to medical ethics it is questionable whether, under well-defined circumstances, it might nevertheless be replaceable by regulatory means. Given the experience of the 'NCT Tissue Bank' that not a single patient refused to donate leftover tissue for subsequent use, a procedure based on contradiction might be regarded as more appropriate. In addition to ethical and legal considerations, the perspectives of those directly involved should be appreciated. We asked our patients, therefore, for their opinions as to whether the established informed consent procedure is satisfactory or in any way deficient.

1 Collecting Data and Tissue for Cancer Research in Heidelberg

For decades, the Department of Urology has focused especially on cancer therapy and research, backed by the fruitful collaboration of the University of Heidelberg, the 'German Cancer Research Center' (DKFZ) and most recently the 'National Center for Tumor Diseases' (NCT). In the early eighties, with the establishment of the Cancer Centre of Heidelberg/Mannheim, a systematic collection of clinical and epidemiological data was begun as a basic requirement of effective cancer research. In 1985, a new database was created to accommodate all available data on patients, past and present, undergoing surgery in the department of urology. This collection directly benefits patients through the 'Heidelberg cancer maintenance program', which reminds patients to attend control examinations.

Besides data, the Institute of Pathology regularly collects routine tissue samples embedded in paraffin, and has compiled a collection going back thirty years. This is remarkable because a minimum period for storage has not yet been legally defined and the 'German Association of Pathology' recommends only three to five years. The resulting paraffin archive contains more than 6,000,000 tissue-blocks. In addition, 5,000 cryo-preserved tissue samples have been collected on an irregular basis since the early nineties, when new methods of analysis were developed. Affiliated institutes collected a further 2,000 samples, and in 2005, after the foundation of the NCT, started the systematic banking of 1,500 more specimens. Therefore, the cancer tissue biobank in Heidelberg contains more than 6,000,000 paraffin- and about 8,500 cryo-samples. Since the year 2000 the urologic tissue blocks have been supplemented by blood and urine samples.

In 2003, the NCT was founded in order to improve interdisciplinary clinical management of cancer patients and as a common platform for cancer research at the University of Heidelberg. This "centre approach" is inspired by the model of comprehensive cancer centres in the United States and involves the Medical Faculty of the University of Heidelberg, its affiliated Hospitals for Thoracic Surgery and for Orthopaedics, the DKFZ and the 'German Cancer Aid'.

As an interdisciplinary institute, the NCT was able to coordinate the efforts of various medical specialities to facilitate standardized data documentation and tissue collection in a single biobank-project. A proposal (207/2005) approved by the Ethics Committee of the Medical Faculty of Heidelberg resulted in the establishment of the biobank in 2005. To guarantee legal security and respect for ethical considerations, the NCT used a procedure attended by a physician, which provided both oral and written information to obtain a one-time, general informed consent from the participant.

In everyday routine there are apparent tensions between the optimised informed consent procedure and prevailing resources. Education regarding tissue preservation conflicts with other far more clinically important issues such as preparation of patients for surgical removal of tumours. Preoperative examinations and information about the intervention procedures are crucial for patients and physicians alike. Patients asked to donate leftover cancer tissue to the biobank usually respond by saying: "If it is for research, why not?" They rarely have further questions and the majority of patients do not even look at the document they sign.

Given that from its inception not a single patient has refused to take part in the 'NCT Tissue Bank', a consent procedure based on contradiction might be more appropriate than the requirement of a statement of informed consent. On the other hand the idea of 'informed consent' has become indispensable to medical ethics and it is questionable whether even in well-defined circumstances it might be replaceable by regulatory means. In addition to ethical and legal considerations, the judgment of those directly involved should be appreciated. Accordingly, we conducted a survey on our patients to evaluate whether the established procedure of informed consent for the collection of removed cancer tissue is satisfactory or inefficient and unnecessary.

2 Methods

In September 2007, we began to ask cancer patients undergoing surgery to answer a questionnaire attached to their participation in the 'NCT Tissue Bank'. The Ethics Committee of the Medical Faculty of Heidelberg approved the study (S-212/2007) and required informed consent in writing from every patient. The preliminary data, from January 2008, are presented here.

The self-assessment form begins by asking for specified characteristics of the patient, which correlate with certain beliefs and judgments: age, sex, language, religion, education, profession, and family status. It then requires a medical history including the nature of the cancer being treated, the certainty of the diagnosis and whether the patient has other malignant or chronic diseases. The HADS-D questionnaire, the German version of the 'Hospital Anxiety and Depression Scale' for measuring psychological reaction to somatic diseases (Zigmond and Snaith 1983), is the only validated instrument applied. Following this general information, the form sets out 34 statements concerning various matters including regulation of the biobank to be rated from 'absolutely true' (1) to 'absolutely false' (6) with an ordinal graduation from 1 to 6. To ease readability and give an overall impression of trends, most of the results are shown as box plots. A larger patient sample size will enrich statistical analysis of the correlation of patient characteristics.

3 Results

First, we tried to determine how well the informed consent procedure was realised in various institutions. The results ranged from non-implementation in the 'Department of General Surgery' (0%), to a positive response from nearly every patient in the 'Clinic for Thoracic Diseases' (100%). The rate of participation in the 'Ear, Nose and Throat Department' was 90% and the 'Department of Urology' 33%, which equals 356 out of 1069 eligible urological cases from October 2005 to October 2007.

When, during the stated time frame, we asked 80 patients to take part in the 'NCT Tissue Bank', none of them refused. The questionnaire circulated simultaneously and had a return rate of 85% (n=68). The patients had a mean age of 60,9 years (32 to 78 years) and 13% were female. The first language of 88,2% was German. 69,2% referred to themselves as religious, 69,1% were of Christian confession. A stable partnership was reported in 82,3% of the cases, although 32,4% did not have children (the mean was 1,35 children per patient). 34,3% had higher education and 44,1% were retired. Tumour diagnoses covered prostate cancer (70,1%), renal cell cancer (22,4%), urothelial cell cancer (4,5%) and testicular cancer (3%). Further, 35,8% had close relatives or friends suffering from cancer.

The results of the HADS-D questionnaire on current mental health (Figure 1) showed that 9 patients had pathological scores related to anxiety or depression (13,2%). 13 had intermediate scores (19%), whereas anxiety seems to be most prevalent with significant scores of 17 patients (25%).

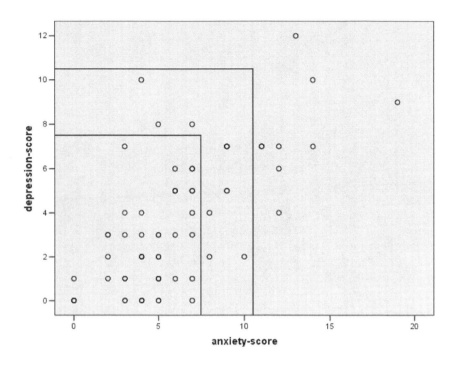

Figure 1: Anxiety and depression scores (n=68). Values lower than 8 are defined unsuspicious. Scores above 10 indicate a pathologic psychical situation

3.1 Right to Withdraw Consent

A large majority of the patients indicated that they would also have given their consent even if a right to withdraw had not been provided (Figure 2). Although the right to withdraw consent at any time and without justification is regarded as a basic requirement for participation in medical research, it didn't seem to be very important in the context of donating leftover cancer tissue. A patient's decision to withdraw would only imply the abandonment of previously collected personal data in most cases.

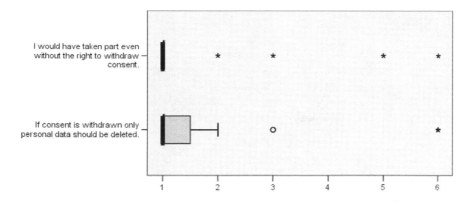

Figure 2: Possibility to withdraw consent and possible consequences (n=47)

Opinion as to the destruction of previously collected tissue was less unanimous (Figure 3). A two-thirds majority (67,4%) opposed the destruction of tissue samples upon withdrawal of consent, indicating a split in the general opinion. About one third (30,4%) thought the tissue ought to be discarded as a consequence of a withdrawal. The low importance attributed to the right to withdraw consent should be taken into account in order not to misinterpret this result.

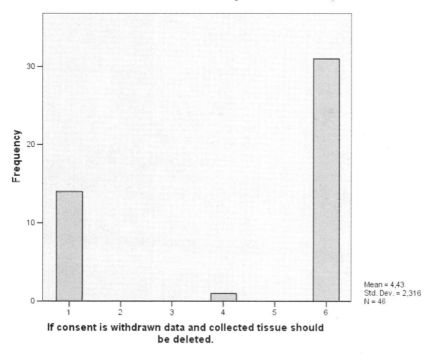

Figure 3: Divided opinion concerning deletion of already collected tissue (n=46)

3.2 Motivation for Taking Part

All of the patients were hopeful that research would find a cure for their disease, and most of them believed strongly in such a development (Figure 4). Personal benefit was not the only motive for taking part, however; altruism was also a strong reason. Circumstantial pressure to participate was not a relevant factor, as a failure to join the biobank held no fear of adverse conditions of treatment.

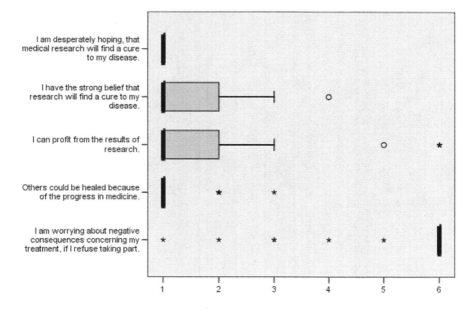

Figure 4: Altruistic intention and hope for personal health benefit are the strongest motivation; negative pressure is not felt (n=59)

In addition, there was a widely felt moral duty to support the progress of research in medicine by delivering good health care services in Germany (Figure 5). It was almost completely understood that project participation was voluntary and that no one should feel forced to take part. Another important criterion was the institute in charge, and the involvement of the University Clinic of Heidelberg seemed to be a positive factor for several patients. A majority said they would have decided against participation, had the biobank been run by industry or a private company.

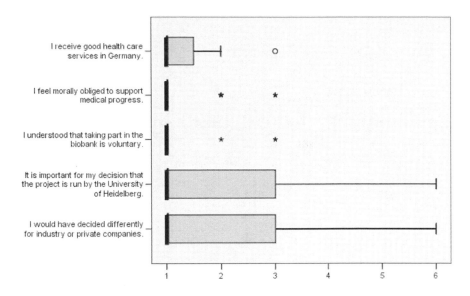

Figure 5: Moral obligation for supporting medical research is realized, but its degree might depend on the institution in charge (n=59)

3.3 The Informed Consent Procedure

Most patients in the study felt that their consent to the storage and examination of tissue samples should be mandatory, although this belief was not absolute (Figure 6). The established informed consent procedure was not considered too protracted, although they wanted to avoid a more detailed decision-making process regarding whether or not to take part. According to their self-assessments, nearly every patient had read and understood the information leaflet and had been satisfactorily informed of the issues. Finally, there was wide acceptance of the use of old tissue samples collected during a time when no consent was required from donors. Whether this practice should be considered as inconsistent with the binding character of consent to new donations will be the subject of further discussion.

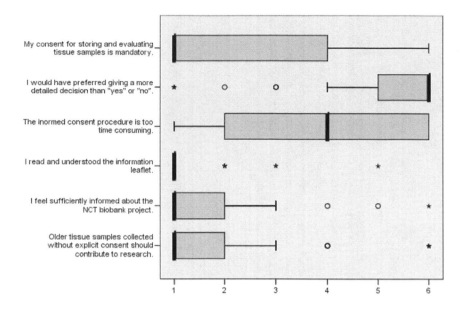

Figure 6: Evaluation of the informed consent procedure (n=57)

3.4 Regulations for Restriction of Usage

Restriction of the terms of consent to certain types of research is a possibility, although a majority of patients did not call for that (Figure 7). As genetic diagnostics received much public interest and criticism in Germany, we evaluated whether this issue was a case for certain restrictions. But genetic investigation of cancer tissue was not problematic for patients, as the majority did not regard the genetic characteristics of cancer cells to be private property.

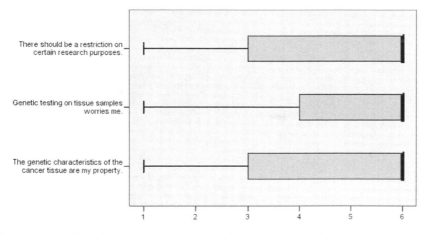

Figure 7: Genetic testing seems less worrisome than expected (n=59)

3.5 Financial Regulation, Benefit Sharing and Feedback

The study also indicated the patients' opinion on compensation. Patients largely agreed that exchange of tissue samples should be free of charge and that donors should receive no financial compensation (Figure 8). They also felt that the profits arising from research, including patents generated from biobank data, should be used for the benefit of the community. Most patients wanted to have feedback on the tissue examination, although this information could have negative consequences for the donor, too.

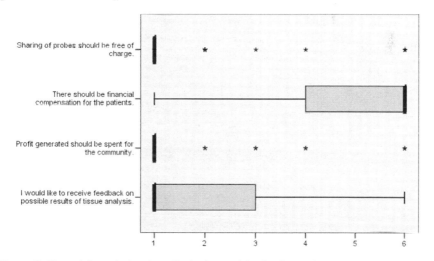

Figure 8: Financial regulation, benefit sharing and feedback (n=59)

3.6 Collected Tissue

Although these questions have not been fully answered, the general attitude toward cancer tissue is that it ought to be defined as a foreign entity, and not as a part of one's own body (Figure 9). Therefore, the use of leftover cancer tissue, additional blood and urine samples is not problematic. Collections of healthy human tissue might be assessed in a different way.

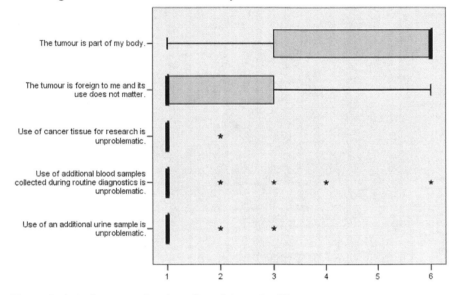

Figure 9: Attitudes concerning the collected tissue (n=56)

4 Discussion

The design of the framework for the collection of bio-specimen resources was the result of the application of profound reasoning on the basis of empirical data. At least 30 empirical studies have investigated views of more than 33,000 people from various backgrounds (Wendler 2006). Over 80% demonstrated a willingness to donate, and those who were unwilling "tended to be concerned with the method of obtaining samples, not the possible use of the samples for research" (Wendler 2006: 545). This statement exemplifies that very little data is available concerning the use of leftover tissue; only five studies are concerned with samples that do not involve an additional act of acquisition (Wendler 2006). To our knowledge, only two empirical publications contribute to the special issue of surgically removed cancer tissue (Malone et al 2002, Start et al 1996). Evidently, the type of tissue collected is highly relevant to possible donors (Barr 2006).

Therefore, our patients' opinions on the informed consent procedure could help clarify our views and lead to adjustments in the regulations. Individuals undergoing the informed consent procedure are in a special situation of anxiety and stress, making them vulnerable and in need of protection. Although the right to withdraw consent did not seem to be of particular importance to them, it should be retained for ethical and legal reasons. Nevertheless, its low overall priority renders the deletion of personal data only possible. The requirement to keep tissue samples despite a withdrawal of consent, while ensuring a high degree of reliability for collected samples, inevitably creates the impression of irrevocability.

Motivation for participation is driven by confidence in medical progress and beneficial effects for the patient's course of treatment as well as for others suffering from the same disease. A significant public benefit that can be achieved quite easily means a strong moral obligation for joining the project; the provision of leftover tissue to non-commercial research institutions, especially in a solidly supported health care system, could be judged as a reasonable return. The right to contradict has, of course, to be assured. It appears to be of great importance that delicate projects like biobanks are conducted by trustworthy non-profit organizations (Jack and Womack 2003).

Surprisingly, genetic testing seems to be less worrisome to the patients than it might have been assumed, considering the German public discourse. The potential danger is intimately connected with efficient data protection: as long as genetic data cannot be associated with real persons there is no need to worry. Only very few patients would like to restrict research to certain permitted purposes; freedom from limitation to specified research projects increases the research potential of biobanks.

Moreover, the actual terms of financial regulation accord with the views of patients (Godard et al 2003). A clearly altruistic weighting has to be emphasised, as no financial compensation for participating patients is demanded, but social benefit sharing is encouraged. Uncertainty prevails about possible personal feedback, which is excluded in the 'NCT Tissue Bank' for mainly practical reasons. In smaller disease oriented projects personal requests can be complied with.

Our data shows that the cancerous tissue and its characteristics are alien to the patient: they are not felt to be a part of the body to begin with and are not regarded as their property. Consequently the patients are unconcerned about disposal of the tissue, and do not worry much about genetic testing in the context of data protection. The results are identical for blood and urine samples although patients are sometimes concerned about the requirement of additional blood samples, and the effect that this might have prior to surgery. Almost every patient is satisfied by an explanation indicating that the amount of blood loss is medically inconsequential. So far, the established framework appears to meet most of the patients' demands. This also holds true for the informed consent procedure. A majority of patients regards their consent as necessary and only wants to decide whether to take part. There is a tendency to the feeling that the information procedure is too extensive;

patients want to be asked to provide consent, but their need to be completely informed is less well developed.

At the same time, practical implementation of informed consent is crucial in that it affects not only patient satisfaction and legal security but also practicability and efficiency of data and tissue collection. The success of tissue collection depends on the motivation and opportunity of physicians and on the prevalence of necessary resources. If the highest standards of consent are to be assured, and the wishes of patients to donate adequately recognized, new resources or organizational structures will be needed (Wheeler et al 2007). Implementation of technical regulations requires infrastructural means, as illustrated by varying degrees of realisation in various departments. By far, not all of the patients suitable for the project are asked to give their consent, although virtually everyone is willing to contribute.

Currently, an informed consent procedure for the donation of human tissue seems to be state-of-the-art for biobanks, although a discussion still prevails about the validity of one-time general consent (Furness 2006) and the necessary degree of information (Moutel et al 2001). Although some empirical data exists, most of the results are inapplicable here due to the special setting. A qualitative distinction can be made in biobanks that are designed exclusively for leftover tissue generated by operative interventions. In that case legitimacy on the basis of valid informed consent is presupposed. Given the high priority placed on consent to planned surgical procedures, an administratively efficient solution would be to include a statement addressing the intended use of leftover tissue in the written form of consent.

Can we identify criteria for evaluating the necessity of an additional formal informed consent? The prevailing risks for participants are surely an important factor. As no additional measures are used to gain tissue samples, and correct coding as well as data protection ensure the patients' privacy, the risk profile seems very moderate. Moreover, supervision and control by an ethics committee ensures that the project is properly designed and avoids possible abuse. Further, the achievement of high prospects for social and individual benefit requires relatively low effort, as leftover material is supposed to be used instead of being thrown away. This constellation creates a moral commitment in patients, which is reflected by a personally felt obligation and an extremely high willingness to participate. Finally, the pre-existence of an informed consent procedure for invasive treatment provides an efficient mechanism by which to realise the assignment of leftover material.

On the other hand, it might be felt that one should ask patients for permission whenever possible; this could demonstrate a balancing of public welfare and respect for autonomy, thus promoting public confidence in medical research. If a separate consent for donation of leftover cancer tissue is ethically indispensable, more personnel will be required to fulfil the high demands of this procedure. Physicians would have to be backed up by assistants, such as e.g. specially trained nursing staff (Wheeler et al 2007), in order to increase rates of inclusion while keeping ethical and juristic standards high.

An alternative might be to generalize the extremely high rate of consent. One could decide to presuppose consent, while enabling contradiction to take place. There is no clear evidence from our data that patients would favour an approach based on contradiction, but at least there does not seem to be decided refusal. The main motive for taking part is altruistic and the patients feel a moral obligation to contribute to research. Wouldn't it presumably be in their interest to enable high inclusion rates in order to achieve medical progress? Most patients feel that their consent should be asked for, but at the same time there is wide acceptance of the use of older tissue samples without explicit agreement. Regulations must meet not only the needs of possible participants and ethical principles, but everyday conditions as well.

The absence of legal security in Germany is the main barrier to a solution. There is no general legislation regulating even the very central issues of biobanking. Therefore, these questions are subject to general legal principles leading to contradictory interpretations and ambiguous conclusions. While some authors think residual tissue has been left behind and could therefore be used without additional requirement (Taupitz and Wicklein 2007), others argue the necessity of an explicit assignment of property accompanied by a specific informed consent (Savulescu 2002). Moreover, regulatory regimes that govern public institutions collecting residual tissue and accompanying data vary between federal states and depend on local ethics committees of the respective medical faculty. This leads to variation in standards of informed consent and ultimately divergent regulation of the use of tissue and data, rendering the exchange of collected material nearly impossible.

Aware of these pitfalls, the 'NCT Tissue Bank' tries to maintain the highest standards. Patients are informed orally by their attending physician and sign a standardized informed consent sheet approved by the local ethics committee. In addition, a patient agreement to use residual tissue and associated data is contained within the contract of treatment signed upon attending at the hospital. This has also been approved by the ethics committee, but is regarded as merely accessory and does not replace oral and written informed consent.

The duty of documentation and monitoring is also unclear. The formal establishment of an informed consent procedure and reliable realisation confirmed by random evaluations have so far been held to be sufficient. It is not considered a duty of the 'NCT Tissue Bank' to archive the consent forms or to check every single case. Failure of a random control to prove given consent could be considered an exceptional error. If the practical result of implementation of such a framework is the use of tissue without consent, the basic idea of informed consent is weakened. But if consent to tissue collection is absolutely mandatory, and too few patients are asked to give it, the statistical power of biobank research would be endangered and could, for ethical reasons (Harris 2005), end in the abandonment of informed consent for the scientific use of left-over cancer tissue (Savulescu 2002;

van Diest 2002). Sometimes less elaborate ethical requirements may prove to be more ethical in a wider context.

Irrespective of the details of ethical decision-making, it would be useful to develop a generally accepted legal framework for biobanks. Regulations for the collection of bio-specimen resources should be arranged in a hierarchy of ambition, depending on the type of tissue, the mode of extraction and the research purpose. Legal harmonisation throughout Germany – or even better Europe – would enable fruitful research co-operation.

References

Barr M (2006) 'I'm not really read up on genetics': biobanks and the social context of informed consent. BioSocieties 1: 251-262

van Diest PJ (2002) No consent should be needed for using leftover body material for scientific purposes. For. BMJ 325: 648-649

Furness PN (2006) One-time general consent for research on biological samples: good idea, but will it happen? BMJ 332: 665

Godard B, Schmidtke J, Cassiman JJ et al (2003) Data storage and DNA banking for biomedical research: informed consent, confidentiality, quality issues, ownership, return of benefits. A professional perspective. Eur J Hum Genet 11: 88-122

Harris J (2005) Scientific research is a moral duty. J Med Ethics 31: 242-248

Jack AL, Womack C (2003) Why surgical patients do not donate tissue for commercial research: review of records. BMJ 327: 262

Malone T, Catalano PJ, O'Dwyer PJ et al. (2002) High rate of consent to bank biologic samples for future research: the Eastern Cooperative Oncology Group experience. J Natl Cancer Inst 94: 769-771

Moutel G, de Montgolfier S, Meningaud JP et al (2001) Bio-libraries and DNA storage: assessment of patient perception of information. Med Law 20: 193-204

Savulescu J (2002) No consent should be needed for using leftover body material for scientific purposes. Against. BMJ 325: 649-651

Start RD, Brown W, Bryant R et al (1996) Ownership and uses of human tissue: does the Nuffield bioethics report accord with opinion of surgical inpatients? BMJ 313: 1366-1368

Taupitz J, Wicklein M (2007) Biobanken: Spannungsfeld zwischen Forschung und Persönlichkeitsschutz. BioFokus 74: 3-7

Wendler D (2006) One-time general consent for research on biological samples. BMJ 332: 544-547

Wheeler J, Agarwal M, Sugden J et al (2007) Experiences from the front-line routine consenting of surplus surgically removed tissue: without investment by the National Health Service fully informed consent for all is not available. J Clin Pathol 60: 351-354

Zigmond AS, Snaith RP (1983) The hospital anxiety and depression scale. Acta Psychiatr Scand 67: 361-370

Legal Issues

What is in a Clause?

A Comparison of Clauses from Population Biobank and Disease Biobank Consent Materials

Susan Wallace, Stephanie Lazor, Bartha Maria Knoppers

Abstract The number of population-based and disease-based biobanks being created for research purposes is increasing. These collections of samples and associated data are being used to discover the links between genes and disease, and the genetic, lifestyle and environmental factors behind common complex diseases. In order to make a decision as to whether or not to provide their consent, potential participants in both types of biobanks need to be informed of the requirements and implications of participation. This comparative examination of clauses contained in consent materials from disease biobanks and population biobanks points to the factors that are specific to each type of biobank and highlights the issues that should be taken into consideration when creating consent materials for biobanking activities.

1 Introduction

Researchers have moved rapidly from searching for single human genes to examining the human genome in detail and comparing genomes with each other. Many are now studying genomic variation across populations, in an effort to discover the links between genes, lifestyle and environmental factors that may cause common complex diseases. It is hoped that these investigations will lead to new knowledge, interventions and therapeutics. In order to realize these ambitions, large resources of high-quality human samples are needed and this has driven the creation of large-scale population-based genetic databases, or population biobanks. A number of these population biobanks are now recruiting participants and beginning scientific study on data and samples. Likewise, the need for a significant store of samples is driving the creation of disease-specific biobanks, where researchers are seeking to link genes with disease. In order to participate in either of these kinds of biobanks, individuals must give their informed consent. Biobanking, and genetic research in general, raises ethical issues that have been much debated, and the determination of appropriate information for inclusion in consent materials for population biobanks has been discussed in depth. An examination of consent materials from disease biobanks may shed light on the information that should be given to individuals to enable them to reach an informed decision regarding participation in population biobank research.

2 Overview

2.1 Definition of a Biobank

Biobanking has been defined as "… the organized collection of biological samples and associated data" (Cambon-Thomsen et al 2007). There has been an increase in the number and types of biobanks. No concise typology exists for biobanks and they could therefore be categorized in many ways. Kelley and colleagues have suggested that this be done according to tissue type (i.e. tumour, DNA), purpose (i.e. research, forensics), ownership (i.e. private, public), volunteer group (i.e. population, disease) and size (i.e. disease group, nation-wide) (Kelley et al 2007). Others group them by function: diagnostic biobanks (i.e. collections in pathology laboratories), therapeutic (i.e. blood banks) and research biobanks (i.e. disease, population) (European Union 2003). Research biobanks can take many forms, from small family-based collections to large-scale epidemiological studies. Attempts have been made to categorize biobanks more definitively (Cambon-Thomsen 2004, Hirtzlin et al 2003), but it is clear that, as they do not follow any one standard format, this will not be an easy task. However, the ability to provide a clear definition of the sort of biobanking project that one is asking people to join would be very useful when approaching prospective participants. Cambon-Thomsen and colleagues see the definition as "… a key element for implementing proper ethical management …. [and] … a prerequisite before starting to inform patients" (Cambon-Thomsen et al 2007). Kaye has called for additional work on the construction of a typology for biobanks (Kaye 2006).

2.2 Population Biobanking Projects

With the completion of the Human Genome Project, population-based epidemiologic research involving genetics increased dramatically. This type of research

> … focus[es] on the prevalence of gene variants in different populations, the burden of diseases, the impact of gene-gene and gene-environment interaction on disease risk, as well as the validity of utility of genetic tests in improving population health (Khoury 2001).

Genetic association studies can point to a link between a gene and disease, but the confirmation of causal connections requires that studies are reproduced; in the past this has proved to be difficult (Smith et al 2005). Large sample sizes can help in the reproduction of results and confirmation of theories. The need for large sample sets and the fact that technologies, such as genetic sequencing, are becoming more advanced and cost effective, are two of the factors that have driven the creation of large-scale population biobanks (Smith et al 2005).

Population biobanks have been defined by the Council of Europe:

> A population biobank is a collection of biological materials that has the following characteristics:
>
> 1. the collection has a population basis;
> 2. it is established, or has been converted, to supply biological materials or data derived therefrom, to multiple future research projects;
> 3. it contains biological materials and associated personal data, which may include or be linked to genealogical, medical and lifestyle data and which may be regularly updated;
> 4. it receives and supplies materials in an organised manner. (Council of Europe 2006).

Population biobanking projects are usually prospective in nature, in that environmental, lifestyle and genetic data and samples are collected from 'healthy volunteers' (chosen randomly from various sources) who represent a certain population. These projects are also usually longitudinal in nature, meaning that participants are studied over a period of time. Populations vary from country-wide, regions within a country, age groups within a country, etc. The collected data and samples also vary, although participants usually donate DNA extracted from blood, urine or saliva samples, provide lifestyle information, such as dietary and smoking habits and allow their bodily measurements, such as weight, height, and blood pressure to be taken. In some projects, participants agree to allow their medical records and/or administrative health records to be linked to their data and samples. These collections of information constitute the resource for future research studies.

Population biobank research most often focuses on illnesses that have wide-ranging effects on a population or explores the genetic diversity of a target population; there is also research being done on what keeps populations healthy. For example, the aim of UK Biobank is

> ... to improve the prevention, diagnosis and treatment of a wide range of illnesses (such as cancer, heart disease, diabetes, dementia, and joint problems) and to promote health throughout society (UK Biobank 2007).

This sense of helping society links population biobanks back to their public health aims. According to Khoury,

> The real promise of genomics and its public health impact will be our improved ability to use genetic information in diagnosing, treating, and preventing ... diseases that we normally do not think of as 'genetic' that are due to complex interactions between multiple genes and the environment (Khoury 2001).

Serving the health of the public is also one of the ways in which population bio-banks have been able to garner public funding and public trust and acceptance. In order to allow researchers to use this resource, however, population biobanks have had to put into place measures to protect the confidentiality of participants. This is most often done by coding[1] individuals' information. The personal data of the participant, such as name, address and telephone number, is replaced with a code that is then used to label biological materials and other information such as completed questionnaires and medical records. The key that enables the data to be reunited with the participant is kept separate, often under different security provisions, so that only a small number of people, usually with a 'duty of care' towards the participant, have access to it. Researchers are given access only to the coded data and/or samples.

The alternative to coding is to anonymize data, by which all identifiers related to the data are permanently stripped, rendering it impossible to reconnect the data with the person who provided it. According to Lowrance, there are several reasons why researchers prefer coding to anonymization of information: to allow data validation or audit, to avoid duplication of cases, to permit the request of additional data if necessary, for checking consent or ethics committee stipulations, to enable a physician or patient to be informed of useful findings and to facilitate research follow-up in future (Lowrance 2002).

Although some biobanks do anonymize data, the current norm is for coded data, as anonymization could be seen to hinder research and may not successfully protect participants' confidentiality (Eriksson and Helgesson 2005).

2.3 Disease-based Biobanking Projects

Disease-based biobanking projects focus on a specific disease or subset of diseases in order to determine their underlying causes and to develop treatments or even a 'cure' for them. Again, biobanking is changing how research is done in this area. The discovery of the gene or genes involved in a disease or condition can open new possibilities for therapeutic strategies. It is not always the case that discovery of the relevant gene or genes leads to a treatment; the gene responsible for cystic fibrosis, for example, was located many years ago but no definitive cure has been found (Zeitlin 2007). It can, however, be an important starting point. As technology has improved, genes can be discovered much more quickly where samples and data from an affected population are available.

---

[1] This paper uses the definitions set out by the International Conference on Harmonisation regarding genomic data (ICH 2007).

Tissue has been collected from individuals for research purposes for many decades (Eiseman et al 2003) and it has been shown that individuals are willing to donate their tissue or blood samples for genetic research (Hamilton et al 2007, Kettis-Lindblad et al 2006). Genomic technologies are now being applied to sample collections as a means of studying multi-factorial diseases. An infectious disease biobank opening in the United Kingdom, for example, seeks "to offer researchers a resource for uncovering those genes that render some people more susceptible to infectious diseases than others" (Towie 2007). A significant amount of work is also being done worldwide to locate the genetic causes of cancer. The Cancer Genome Atlas project, a US project currently in the pilot stage, aims,

> … to determine the feasibility of implementing a full-scale project whose aim would be to develop a complete 'atlas' of the genomic alterations involved in cancer. This compendium of changes could accelerate the development of new targeted approaches to diagnose, treat and prevent cancer that are based on the central feature of cancer, namely, that it is fundamentally a disease of the genome (TCGA 2006).

The creation of new resources for such work, as opposed to the use of existing sample collections, avoids several problems. The informed consent requirements associated with existing collections may not allow them to be used retrospectively without obtaining further consent from participants regarding the additional uses to which their samples might be put. There is also variation in the methods by which samples are collected, processed and stored, the quality of the samples may not be consistent and information about the samples may be limited or non-existent (Eiseman et al 2003). The creation of new collections may therefore be more cost effective and efficient.

In contrast to cancer, an area of study that has traditionally had strong support and funding, other 'rare' diseases are now being studied, in which the number of affected people is very low. It has in the past been difficult to conduct research into rare diseases due to competition for limited research money (Hampton 2006) and because researchers may not see enough patients to collect a sufficient number of samples for their work. In response to these difficulties, organizations advocating for some disease-based patient groups have begun creating their own biobanks (Marcus 2006, Merz et al 2002, Terry et al 2007). By establishing and controlling the resource or biobank, these groups have found that they can initiate research on the particular disease or condition of their members by providing the samples, data and research money that might not otherwise be available. In certain cases, such as PXE International, the founder of the biobank has been named on the application for the gene patent and, through negotiations with researchers have, "… retained authorship of papers and ownership rights of patents to ensure broad and affordable availability of the [diagnostic] test and to retain influence over downstream development" (Merz et al 2002). They hope their efforts will accelerate research and produce therapies for affected individuals.

Disease-based and population biobanking projects vary, then, in many ways. They differ in regard to the numbers and types of individuals that they recruit, the kinds of data and samples collected and the purposes for which the research is being conducted. They are similar in that they both rely upon collections of biological samples and related data to hunt for treatments for diseases and conditions that affect millions of people around the world. They also both require informed consent materials for use in the recruitment of participants for their research projects.

2.4 Informed Consent

The informed consent process is an important component in the ethical conduct of experimental research. One facet of the process is the use of informed consent materials, the traditional means of giving information about participation in a clinical intervention or research project. 'Consent materials' for this study comprise information sheets or pamphlets and an accompanying document for signature by the participant, by which consent is indicated. Although often separate, these two pieces may be combined into one document. Other types of consent materials might include videos and interactive computer programmes (Flory and Emanuel 2004). The structure of informed consent materials – how they should be written and presented – as well as the information they should include, are much debated issues. The latter of these is the focus of this paper.

There has been considerable discussion as to what information needs to be inclued in consent materials to ensure that the individual has been 'informed' (Deschênes et al 2001). Population-based genetic studies raise particular issues including the need for broad consent (as the future uses of samples and data cannot always be foreseen), the protection of confidentiality (through coding and anonymization procedures), and access (determining who is to have access to data and samples, and authorized purposes for it). It is vital that the content of informed consent materials are well thought through and carefully written (Shickle 2006). Given the differences between disease-based and population-based projects, can the information provided to prospective participants in a disease biobank give us clues as to the information that is necessary for participants in a population biobank?

3 Materials and Methods

3.1 Background

Previous work has been done in regard to the elements that are essential for inclusion in consent materials for population-based genetic research studies (Beskow et al 2001). Now that many biobanks are moving from the planning stages into recruitment, we can benefit from their experience. The P3G working group on Ethics, Governance and Public Engagement (IWG3)[2] decided to revisit the issue of informed consent. Recently, in work led by the Policymaking Core[3] of the 'Centre de recherche en droit public' at the Université de Montréal, Montréal, Canada, generic information pamphlets and consent forms were created for use in population biobanking. The process of creating these generic consent materials is described elsewhere (Wallace et al 2008), but the work gave rise to the hypothesis that is discussed in this paper: Are provisions required in the informed consent materials of population biobanks different from those necessary for other types of biobanks? If so, could an examination of the consent materials used by disease biobanks assist us in determining the clauses that should be in population biobank consent materials? To answer these questions, an analysis was carried out that compared clauses in consent materials from disease-specific biobanks with those considered to be important for inclusion in population-based biobank materials.

3.2 Methodology

Informed consent materials were located, by way of internet searches, for ten disease-specific biobanks, three cancer biobanks and one rare disease biobank (referred to collectively as "disease biobanks"). Of the 14 collections, ten were based in the United States and four in Europe. No attempt was made to balance nationalities or types of biobanks; not all projects published their consent materials and therefore only those that were available over the internet were used.

The provisions of the consent materials of these 14 disease biobanks were tabulated under the headings developed by the IWG3 for purposes of its draft generic information pamphlet. These headings are listed in Table 1.[4]

---

[2] IWG3 is a working group of the Public Population Project in Genomics (P^3G), a not-for-profit consortium created to promote collaboration and harmonization between researchers working in population genomics.

[3] Cores are independent research projects that contribute to P^3G activities.

[4] The headings are based on discussions held at the May 2007 meeting of P^3G held in Montreal, Canada. These clauses have been renamed and rearranged in subsequent drafts of the generic pamphlet, but the content remains roughly the same.

Table 1: Clauses included in May 2007 draft P3G generic information pamphlet

Invitation	**Benefits**
• Invitation to join	• Direct individual benefit from participating
• Organization/Support	• Return of examination results
• Funding	• Return of research results
Overview of project	**Project confidentiality**
• Aim	• Data storage
• Number of participants	• Sample storage
• Method of recruitment	**Access by others**
• Duration of project	• Requirement for scientific/ethical review
Governance of project	• Access by researchers
• Oversight	• International/commercial access
• Approval of project	• Return of samples/data to project
Recruitment	•Access for other than research purposes
• Requirements	Commercialization
•Access to medical records	Compensation
• Future research	Withdrawal
Risks	• Withdrawal procedures
• Physical discomfort	• Degrees of withdrawal
• Loss of confidentiality	Thank you / contact information

The clauses in the consent materials were aligned with the relevant headings as a means of comparing and contrasting the information contained in disease biobank and population biobank information pamphlets. A complete and formal analysis of the materials was not conducted; the information points generally to the similarities and differences between them and raises questions for future research. Some of the findings are presented and discussed further in this paper.

It should be noted that the names of the disease biobanks included in the study are not disclosed here, for the primary reason that this research aims to identify the information that they provide in their materials, rather than to criticize any of their decisions. In addition, there is no intention to imply that participants have been inadequately informed about projects, as the authors have no knowledge of the information discussed during the informed consent process, nor of additional materials that may be been distributed. Further, it is unclear whether the internet materials are official or merely examples; the materials actually used in connection with some projects may vary.

4 Discussion

4.1 Invitation and Overview of Project

All of the 14 disease biobanks stated that they were research projects. The aim of the study and how it would proceed were described in depth by all, as was the recruitment strategy. Six of the 14 discussed how long the project would last. Only three biobanks told potential participants that researchers would be restricted in their use of data and samples to a particular disease; only one of those three presented this as an optional choice for potential participants. One would assume that the focus of the research of disease-based biobanks would be specific to the disease in question. Population biobanks are necessarily vague about the research that will be done using their data and samples. They expect to study multi-factorial diseases, such as heart disease or diabetes, but the complexity of these diseases makes it difficult to specify to prospective participants the precise nature of the research that will be using their information. This is a controversial issue (Greely 2007) but such work may nevertheless be necessary. As Greely states,

> Given the still-high cost of collecting samples and phenotypic data and the low costs of genotyping samples, it makes sense to construct genomic databanks that can be used to study a wide range of problems (Greely 2007).

For disease-based biobanks, though, the focus is necessarily on the disease of interest. Advocacy groups have established infrastructures for the collection of data and samples from the affected parties, creating a useful, and in some cases unique, resource for researchers. Time, effort and expense have gone into building these resources, with the goals of adding knowledge, discovering treatments and perhaps providing cures. One would assume, therefore, that these precious resources, sometimes with participants numbering only in the hundreds, would focus on that particular disease and perhaps those related to it. Perhaps researchers and advocates recognise that information is often found in unexpected places and that the imposition of limitations might be detrimental to research objectives. If samples are in limited supply, however, participants might be expected to want them to be used to explore their own disease. Further, explanation as to what the study involves is one of the core requirements of the ethical conduct of research. One would expect consent materials to state whether the donations of participants are being used in the study of a specified disease or diseases, or to create a general resource for the study of various other diseases.

4.2 Recruitment, Risks, Benefits and Confidentiality

All 14 projects explained the processes involved in participation, including, for example, blood draws, questionnaires, and measurements. The risk of discomfort from having blood drawn was mentioned by eight biobanks, while five classified the potential for loss of confidentiality as a risk. All projects discussed how data and samples would be kept confidential. Coding, as with population biobanks, was the preferred method of protecting the confidentiality of information. Nine biobanks stated clearly that the research was not expected to provide any individual benefits to participants. All of the disease banks were very clear that participation was voluntary; they gave ample information on withdrawal procedures and stressed that there would be no penalty for not participating or withdrawing their participation in the future.

The question as to whether participants should receive any individual results from genetic research has been much debated and raises many issues. (Cambon-Thomsen et al. 2005, Knoppers et al. 2006, Pullman and Hodgkinson 2006). Many population biobanks have decided that individual results will not be given to participants, for several reasons. For one, "… results are usually aggregated and considered to be of purely scientific interest" (Cambon-Thomsen et al 2007) as opposed to specific clinical information upon which one might be able to act. Beskow and colleagues note that to deliver individual results might cause participants to confuse researchers with clinicians. They believe this distinction must be kept in place; it is "… the researcher's 'obligation' to participants to conduct good science and disseminate findings widely …" (Beskow et al 2001) but not to provide clinical interventions. From a practical perspective, as large-scale population biobanks are often longitudinal and have tens or hundreds of thousands of participants, it would be very difficult and expensive to be in contact with individuals about any eventual findings. As this is a difficult issue, the consent materials should clearly state whether or not individual results will be provided (Deschênes et al 2001). Participants in many population biobanks may, if they wish, receive results from the measurements taken during the 'joining session'.

One might assume then that disease biobanks, especially those studying rare diseases with small sample sizes, would be in a position to deliver individual results of genetic tests. Affected participants will have already been diagnosed, and would no doubt be expected to be interested in any clinical or therapeutic applications of their information. But this was generally not the case. Seven biobanks specifically stated that no information from genetic tests would be returned to participants (although results from other clinical tests might be provided). Two biobanks said that results would be delivered; three others indicated some sort of results (such as the results of chromosomal tests, particularly if a mutation or deletion was found). The two remaining biobanks made no specific mention of delivery of individual test results, leaving the matter subject to ambiguity. It might be interesting to pursue this issue with participants in disease-specific biobanks in order to clarify the procedures involved and shed light on this issue. Based on these results, it is not clear whether this issue is as controversial for disease bio-banks as it is for population biobanks.

4.3 Governance, Ethics Approval and Funding

Proper governance of a biobank, including the establishment of scientific and ethi-cal review processes and disclosure of funding mechanisms is important for ensur-ing that biobanks have the trust of participants and the public. After all, "... the process of building trust is central to all kinds of biobank projects, whether they rely on patient or on general population studies" (Cambon-Thomsen et al 2007). Nine biobanks cited the sources of their funding. Nine also discussed their organ-izational support, but only four mentioned whether there was oversight of the pro-ject itself and which body provided it. Six stated that the project had been ap-proved by an ethics committee. It may be surprising that more did not mention these undoubtedly reassuring facts; it has been argued, though, that ethics commit-tee approval should not be mentioned in consent materials on grounds that it might mislead prospective participants into believing that the study is 'safe' and that they should therefore take part (Beyleveld and Longley 1998).

4.4 Access by Others

As the main goal of population biobanks is to provide resources for scientific re-search, determining who will have access to data and samples and how that access will be granted are necessarily important issues. It is also vital that the resource, which exists as a result of the altruism of thousands of donors, is used in the inter-est of the participants, in order to maintain this good will (Greely 2007). Custodi-ans of population biobanks must ensure "... that samples are used for the public good and for publicly endorsed ends" and not exploited by commercial interests (Williams and Schroeder 2004). In addition, there are questions regarding the cross-border use of data by international researchers and whether foreign data pro-tection standards are equal to those in the country where the biobank is located (Lowrance 2002). Only five disease biobanks gave researchers from other coun-tries access to data and/or samples, and only four made provision for access by re-searchers from commercial companies. The four disease biobanks that mentioned both issues focussed on cancer and multiple diseases, as opposed to a single dis-ease (other than cancer). Why the other biobanks did not include this information is open to speculation and further research.

What is mentioned is that individuals participating in the research will not gain financially from any products that are developed based on the research, as is the case for population biobanks. Rothstein contends that this is as a result of legal cases where researchers were seen to unjustly benefit financially from participants' data and samples (Rothstein 2005). Perhaps being notified in this way is seen as sufficient to inform prospective disease biobank participants of potential commercial and international interests in their information. Based on the draft generic information pamphlet, these pieces of information are separate in population biobank materials.

5 Conclusions

This short study into consent materials used by disease biobanks began as a way to gather additional information on the clauses that should be included in the informed consent materials of population biobanks. It has also raised questions, however, as to the general information that is needed to inform participants about research projects. As discussed, there are significant similarities between disease and population biobanks, such as the means by which the confidentiality of participants' data and samples is protected, the emphasis on research rather than treatment, and the goal of finding new knowledge, interventions and therapeutics to benefit affected individuals and society in general. It has been demonstrated, however, that there are differences between biobanks that could be examined further. For disease biobanks, the following have been highlighted: access by participants to individual results, the focus of the research and the scarcity of information regarding governance structures and access to data and samples by international researchers and those from commercial interests.

The analysis of this data is not complete. Further investigation is necessary to determine whether the issues raised are of any real importance to disease biobanks. It is also not clear whether the results are reliable, as they may have been confounded by lack of data or misinterpretation of the text of the consent materials. Some issues require greater exploration, including the research uses of data and samples, access by third parties (i.e. insurers, relatives) and oversight mechanisms. Most importantly, further comparative work on biobanks requires the development of a comprehensive typology. Only with a typology in hand, and more detailed study into the issues, can true comparisons be made between biobanks, comparisons that might yield meaningful results and enable informed recommendations to be made in the future.

Acknowledgments The authors wish to acknowledge the work of the members of the International Working Group on Ethics, Governance and Public Engagement of the Public Population Project in Genomics (P3G) Consortium in the creation of these generic tools. The Université de Montréal Policymaking Core is funded by the Canada Research Chair in Law and Medicine, Genome Canada, Génome Québec and the Centre de recherche en droit public, Université de Montréal.

References

Beskow LM, Burke W, Merz JF et al (2001) Informed consent for population-based research involving genetics. JAMA 286: 2315-2321

Beyleveld D, Longley D (1998) Informing potential participants of local research ethics committee approval of research protocols. Med Law Int 3: 209-222

Cambon-Thomsen A (2004) The social and ethical issues of post-genomic human biobanks. Nat Rev Genet 5:866-873

Cambon-Thomsen A, Rial-Sebbag E, Knoppers BM (2007) Trends in ethical and legal frameworks for the use of human biobanks. Eur Respir J 30: 373-382

Cambon-Thomsen A, Sallée C, Rial-Sebbag E et al (2005) Population genetic databases: Is a specific ethical and legal framework necessary? GenEdit 3:1-13 www.humgen.umontreal.ca/int/genedit.cfm?idsel=1312. Accessed 22 February 2008

Council of Europe (2006) Recommendation of the Committee of Ministers to member states on research on biological materials of human origin https://wcd.coe.int/ViewDoc.jsp?id=977859&Site=COE. Accessed 12 September 2007

Deschênes M, Cardinal G, Knoppers BM (2001) Human genetic research, DNA banking and consent: a question of 'form'? Clin Genet 59: 211-239

Eiseman E, Bloom G, Brower J et al (2003) Case Studies of Existing Human Tissue Repositories: "Best Practices" for a Biospecimen Resource for the Genomic and Proteomic Era. RAND, Santa Monica

Eriksson S, Helgesson G (2005) Potential harms, anonymization, and the right to withdraw consent to biobank research. Eur J Hum Genet 13: 1071-1076

European Union (2003) European Union Workshop on Optimisation of Biobanks: Biobanks for Health: Optimising the Use of European Biobanks and Health Registries for Research Relevant to Public Health and Combating Disease. http://www.fhi.no/dav/1F1C30AB2C.pdf. Accessed 12 Sep 2007

Flory J, Emanuel E (2004) Interventions to improve research participants' understanding in informed consent for research: a systematic review. JAMA 292: 1593-1601

Greely HT (2007) The uneasy ethical and legal underpinnings of large-scale genomic biobanks. Annu Rev Genomics Hum Genet 8: 343-364

Hamilton S, Hepper J, Hanby A et al (2007) Consent gained from patients after breast surgery for the use of surplus tissue in research: an exploration. J Med Ethics 33: 229-233

Hampton T (2006) Rare Disease Research Gets Boost. JAMA 295: 2836-2838

Hirtzlin I, Dubreuil C, Preaubert N et al (2003) An empirical survey on biobanking of human genetic material and data in six EU countries. Eur J Hum Genet 11: 475-488

ICH (2007) International Conference on Harmonisation Definitions for Genomic Biomarkers, Pharmacogenomics, Pharmacogenetics, Genomic Data and Sample Coding Categories (E15). http://www.ich.org/LOB/media/MEDIA3383.pdf. Accessed 20 February 2008

Kaye J (2006) Do we need a uniform regulatory system for biobanks across Europe? Eur J Hum Genet 14: 245-248

Kelley K, Stone C, Manning A et al (2007) Population-based biobanks and genetics research in Connecticut. http://www.ct.gov/dph/LIB/dph/genomics/BiobanksPolicyBrief.pdf. Accessed 21 February 2008

Kettis-Lindblad A, Ring L, Viberth E et al (2006) Genetic research and donation of tissue samples to biobanks. What do potential sample donors in the Swedish general public think? Eur J Public Health 16: 433-440

Khoury MJ (2001) Informed consent for population research involving genetics: a public health perspective. http://www.cdc.gov/genomics/population/publications/editorial.htm. Accessed 7 September 2007

Knoppers BM, Joly Y, Simard J et al (2006) The emergence of an ethical duty to disclose genetic research results: international perspectives. Eur J Hum Genet 14: 1322-1322

Lowrance WW (2002) Learning from Experience: Privacy and the Secondary Use of Data in Health Research. The Nuffield Trust, London

Marcus AD (2006) Patients with rare diseases work to jump-start research. The Wall Street Journal Online 11 July 2006

Merz JF, Magnus D, Cho MK et al (2002) Protecting subjects' interests in genetics research. Am J Hum Genet 70: 965-971

Pullman D, Hodgkinson K (2006) Genetic knowledge and moral responsibility: ambiguity at the interface of genetic research and clinical practice. Clin Genet 69: 199-203

Rothstein MA (2005) Expanding the ethical analysis of biobanks. J Law Med Ethics 33: 89 – 101

Shickle D (2006) The consent problem within DNA biobanks. Stud Hist Philos Sci Part C: Stud Hist Philos Biol Biomed Sci 37: 503-519

Smith GD, Ebrahim S, Lewis S et al (2005) Genetic epidemiology and public health: hope, hype and future prospects. Lancet 366: 1484-1498

TCGA (2006) The Cancer Genome Atlas Data Release Workshop Summary Report. http://cancergenome.nih.gov/components/TCGA_101706.pdf. Accessed 20 February 2008

Terry SF, Terry PF, Rauen KA et al (2007) Advocacy groups as research organizations: the PXE International example. Nat Rev Genet 8: 157-164

Towie N (2007) London hospital launches infectious disease 'biobank'. Nat Med 13:653

UK Biobank (2007) UK Biobank Information Leaflet. www.ukbiobank.ac.uk/docs/infoleaflet 0607.pdf. Accessed 13 Sep 2007

Wallace S, Lazor S, Knoppers BM (2008) Consent and population genomics: The creation of generic tools. Submitted

Williams G, Schroeder D (2004) Human genetic banking: altruism, benefit and consent. New Genet Soc 23: 89-103

Zeitlin PL (2007) Emerging drug treatments for cystic fibrosis. Expert Opin Emerg Drugs 12: 329-336

Informed Consent to Collect, Store and Use Human Biological Materials for Research Purposes

An International Framework

Mariaelena Salvaterra

Abstract The increased buildup of human biobanks developed out a need for re-thinking traditional research ethics. In particular, the advancement of research using samples of human origin (namely body fluids, cells, tissues, intra-cellular substance, DNA) calls for the definition of a new model of informed consent appropriate to biobank research.

Even though some recent studies doubt the importance attributed to informed consent, the international debate on ethical and legal aspects of research using human biospecimens and associated data focuses on consent as the *priority issue* of biobanking research.

This paper contains a review of the international ethical and legal framework of consent requisites for retrospective and prospective research using biobank samples. Critical terms and definitions are examined to show that different standards and recommendations concerning informed consent are present at regional, national and international levels. Additionally, different contexts in which a waiver of informed consent can be accepted are discussed and the question of criteria for future regulations is raised. The ultimate aim of this paper is to highlight that there is no unified recommendation concerning the type of consent that should be sought among patients interested in biobanking research at present. Accordingly, the review concludes by calling for a specific biobanking research ethics to deal with the meta-ethical questions raised by informed consent appropriately.

1 Introduction

During the last decade, the constitution of human biobanks came out of a need for re-thinking traditional research ethics (Knoppers 2005). In particular, the accelerated development of research using samples of human origin – body fluids, cells, tissues, intra-cellular substance, DNA – called for the definition of a new model of informed consent appropriate for biobank research.

Although some recent studies doubt the importance attributed to informed consent (Hoeyer et al. 2005), the international debate on ethical and legal aspects of research using human biospecimens and/ or associated data focuses on consent as one of the priority issues of biobanking research.

Many scholars have noted (Deschenes et al. 2001) that it is the difficulty (and often the impracticability) to obtain a truly informed consent that makes informed consent for biobanking research so challenging. This is due to the fact that specificities of future research are often not known at the time of consent request.

Focusing on the issue of so-called *secondary uses*, this article provides a review of the international framework of consent requisites for research using biobank samples.

This analysis aims at highlighting that no unified recommendation on the type of consent that should be sought among patients/ donors participating in biobanking research exists at present. Instead, there is an international ethical and legal patchwork that pools strict (specific informed consent) and less strict regulations (presumed consent).

This fragmentation concerns scientists, ethicists, lawyers and policy makers and gives rise to a common recommendation to unify current regulations and consent models for biobanking research (Maschke 2005).

2 Materials and Methods

The analysis described in this paper was based on the review of: a) ethical and legal documents (recommendations, guidelines, laws) on biobanks promulgated at national, regional and international levels (table 1); b) articles (also called "literature") on ethical and legal issues surrounding biobanking research.

First, documents and articles were found via Internet, using both general (for example, google) and specific (for example, university websites) search engines.

Key words utilised to search for the mentioned materials were: "biobank", "biobanking research", "biobank ethics" and "informed consent for biobanking".

Additionally, materials used for building the ethical and legal framework of consent for biobanking research, as it is described in this paper, were found in the references of the gathered documents.

Subsequently, the materials were analysed focusing on ethical and legal issues at stake in the international debate on biobanking research.

3 Results

The review of the documents mentioned earlier led to an ethical and legal patchwork of biobanking research, in which concepts and regulations concerning biobanks are far from a shared and integrated system at international, regional and national levels.

3.1 Confused (and Confusable) Words

The review of ethical and legal documents, as well as the analysis of literature, showed that there are currently no uniform definitions of the term "biobank". Published in PubMed for the first time in 1996 (Loft and Poulsen 1996) the concept of "biobank" is commonly used to refer to different collections of human biological materials and/ or associated data.

Current concepts comprise the following pairs of biobank definitions:

a)   individual and population biobanks
b)   biobanks of data and biobanks of samples and data associated with them
c)   biobanks of samples permanently preserved and biobanks of samples temporarily or permanently preserved
d)   DNA biobanks and tissue biobanks

Besides the concept of biobanks, other confused and confusable words refer to the identifiability of samples and data.

Concerning the levels of identification of biospecimens and information, the categories commonly used are the following: 1) identifiable/ unidentifiable materials, 2) identified/ unidentified materials.

As reported in the European Recommendation concerning research on biological materials of human origin, "identifiable biological materials are those (...) which, alone or in combination with associated data, allow the identification of the persons concerned either directly or with the use of a code" (art.13, i.). In the latter case, the coded materials are also referred to as "linked anonymised materials" (Council of Europe 2006).

"Non-identifiable biological materials, also referred to as "unlinked anonymised materials", are those (...) which, alone or in combination with associated data, do not allow, with reasonable efforts, the identification of the persons concerned" (art. 13, ii.).

As described in most American regulations and guidelines (Thomas 2006), *unidentified samples*, sometimes termed "anonymous", are those provided by repositories and taken from a collection of unidentified human biological specimens.

Unlinked samples, sometimes termed "anonymised", are those lacking identifiers or codes that can link a particular sample to an identified specimen or a particular human being.

Coded samples, sometimes termed "linked" or "identifiable", are the ones taken from a collection of identified specimens with a code rather than personally identifying information such as name or social security number.

Identified samples, are those supplied by repositories from identified specimens with a personal identifier allowing the researcher to link the biological information derived from the research directly to the individual from whom the material was obtained.

3.2 Contentious Bioethical Issues: the "Hard Hoof" of Informed Vonsent in the Maze of "Hard" and "Soft Law" Agreements

The review of laws, regulations, guidelines and ethical statements on biobanks also revealed that a number of ethical and legal issues on biobanking research presently remain controversial: informed consent, privacy, return of results, governance structure, public involvement, commercialisation and benefit sharing.

Besides the *cool* issues – as governance structure and public involvement (namely, citizenship consultation) – the "old matter" of the informed consent keeps to play a crucial role in biobanking debate.

What remains extremely controversial on the international level is the type of informed consent that should be sought for in biobanking research; more precisely, the kind of information that should be given to patients/ donors and the type of consent that should be requested in the light of the provided information.

As well as conceptual issues, consent requirements for biobanking research are not uniformly regulated. Indeed, international, European and national laws and regulations concerning informed consent in the context of using biospecimens and/ or associated data for research purposes are far from a common consensus.

In particular, the fragmented character of ethical and legal documents becomes clear when it comes to the topic of collecting and storing biosamples for future research uses, also called *secondary uses*.

3.2.1 Informed consent for retrospective biobank research

As far as research using archived biosamples and/ or associated data, retrieved from medical care setting – also termed *retrospective biobank research* – is concerned, the general trend in super-national (international and regional) regulations is a waiver of informed consent when biospecimens and information are unlinked, anonymised and when certain conditions are met.

In its report on genetic data-bases, the World Health Organisation permits the use of such materials when made anonymous in such a way that the identification of a "sample source" is not possible (World Health Organisation 1998).

The Human Genome Organisation (Human Genome Organisation 1998), the Council for International Organisations of Medical Sciences (Council for International Organisations of Medical Sciences 2002) and the UNESCO (United Nations Educational, Scientific and Cultural Organisation 1997) have adopted a less restrictive approach, allowing that stored samples and data are used not only in unlinked anonymous, but also in coded form without a re-consent for research.

Moreover, these documents demand that each research project is approved by an ethical review committee.

Similarly, the European Recommendation concerning the research use of human biological materials (Council of Europe 2006) permit the waiver of informed consent for research using previously stored materials if they are unidentifiable and on condition that "such use does not violate any restrictions placed by the person concerned before to the anonymisation of materials" (art. 3).

Despite the common trend accepted at international and regional levels, national regulations contain different recommendations concerning informed consent requirements for research with materials that have already been stored and/ or associated data.

Policies in Canada, Germany, Norway, the Netherlands and the United States permit the use of stored samples without consent if the samples are not identifiable. Iceland leaves the authority over making decisions on the need for a new consent for research using stored samples to its National Bioethics Commission.

In Estonia and the UK consent is not required for further use of collected samples due to the fact that individuals give a broad consent to research carried out with their samples at the time of biospecimens collection (Knoppers 2005).

3.2.2. Informed consent for prospective biobank research

Apart from some shared principles, which mainly concern formal aspects of informed consent, regulations – namely, laws, guidelines, ethical recommendations – concerning consent requisites for research using (or re-using) human biosamples and/ or associated information (also called *prospective biobank research*), vary significantly at international, regional and national levels.

The core of common rules refers to the following provisions: consent should be free, explicit, prior to any proposed research project and given on the basis of a mindful choice.

These requisites comply with traditional biomedical ethics which focuses on the expressed, informed consent of the patient/ research subject as a safeguard of individual autonomy.

On the contrary, information on the research use of samples and/ or data as well as on the related consent form are both regulated in a different way at supernational and national levels.

As it is shown in table 1, regulations concerning consent models for biobanking research range from broad (also called unrestricted) to specific (also called fully restricted) consent (Box 1).

At the European level, both legislations and recommendations tend to require a consent form that is as specific as possible.

While the Convention for the protection of human rights and dignity of the human beings (→ is this the official name of the convention? If yes, shouldn't it be capitalized?) states, with regard to the application of biology and medicine, that "consent for using body parts for purposes other than that for which they were originally removed should be appropriate according to national laws" (art.22), in the record of the convention it says that consent for such uses should be specific (Council of Europe 2005).

In the same way, the recommendation concerning research on biological materials of human origin requires a specific consent for any foreseeable research use and a consent form as specific as possible for undefined/ unplanned research studies (Council of Europe 2006).

On the national level, the fragmentation of prospective biobanking research regulations and consent models remains at stake (→ what exactly is meant here?).

On the one hand, every country names different provisions for consent requisites for biobank research. On the other hand, some countries endorse conflicting consent models.

In the United States, research involving the use of tissues samples and other human biological materials is regulated by both state and federal laws (Merz 2003).

Concerning federal laws, the so-called "Common Rule" – a series of federal regulations concerning research studies funded by federal departments and agencies – calls for voluntary informed consent and oversight of each research protocol and consent process by a local institutional review board (IRB).

The Common Rule permits a waiver of informed consent when samples and/ or data are unidentifiable and the IRB acknowledges risks for subjects to be low. The Common Rule also permits a waiver of informed consent requirements in the case of identifiable materials when the IRB acknowledges that research causes minimal risks, respects personal rights and that it would be impractical to obtain informed consent.

Although the Common Rule only applies to research studies funded by the participating federal agencies, several states have laws demanding that researchers adhere to the Common Rule, regardless of whether the research is funded by a private commercial sponsor or federal funds are used.

More serious are the provisions mentioned in the National Bioethics Advisory Commission Report titled "Research involving human biological materials: ethical issues and policy guidance" (National Bioethics Advisory Commission 1999).

The NBAC Report recommends that research consent forms should leave several options open to potential subjects, ranging from complete refusal of the use of samples for research to a series of limited permissions and permitting coded use of their materials for any kind of future study (multi-layered consent).

The NBAC Report also recommends an improvement of the informed consent process, including a separation of obtaining consent to research use of human biological materials from obtaining informed consent to clinical procedures.

In European (Germany, Iceland, United Kingdom) and extra-European countries (Switzerland, Estonia, Japan, Latvia) the consent model mostly recommended is the broad consent.

Several countries do not clarify the type of consent that should be sought for in biobank research, but limit their recommendations to informed and expressed consent (Denmark, Netherlands, Spain, Norway). Other countries explicitly recommend the requirement of specific informed consent – Italy, France, Sweden (Bernice and Caplan 2006). Some countries (Australia) state that consent for future use of data and tissue in research may be specific, extended or unspecified.

A common and widespread trend in the countries mentioned is a waiver of informed consent requirements concerning biobank research when biosamples and/ or associated data are unidentifiable and when a review by an ethics committee is granted for all research studies using biological materials and/ or related information.

4 Discussion

The review of regulations – of both "hard" and "soft" regulations – and literature on biobanking research has shown that there is currently no unified recommendation concerning biobank concepts and norms.

Instead, there is a strong tendency to define and regulate research using human biospecimens and/ or associated data in different ways at international, regional and local levels.

In particular, the review highlighted that one of the most controversial issues surrounding biobanking is the definition of consent requisites for the collection, storage and future use of biosamples and/ or associated information for research purposes.

Indeed, while the request of a consent form for banking and the use of biomaterials for research is acknowledged internationally, the definition of consent requirements for biobanking is not regulated uniformly.

With respect to that, international, regional and national regulations provide different rules for retrospective and prospective research with human biomaterials and/ or linked data.

While retrospective biobank research is governed by a general rule saying that a waiver of informed consent when biosamples and/ or data are anonymous and other conditions are met, research using biospecimens and/ or information for immediate or future unforeseeable uses is regulated by a patchwork of ethical and legal provisions at international, regional and local levels.

Whereas American and European legislations and "soft law agreements" tend to request a specific consent form – at least for identified or identifiable biomaterials and/ or data – national laws and regulations tend to apply a broad consent on condition that biosamples and/ or data are anonymous, the research project is approved by an ethics committee (or other competent body) and the "opt-out" of the patient/ research subject is assured.

The most important observation made in the course of this analysis (which is far from being exhaustive) is the widespread approach to reject traditional research ethics (and the traditional informed consent) in favour of a general authorization given by patients/ research subjects to use (or re-use) their materials and/ or data for research purposes.

Concerning this aspect, there is no difference between retrospective and prospective research. In both cases, the tendency is to take down the high level of human right protection assured by the traditional informed consent – which requires a specific information and authorisation – in order to apply a broad consent, permitting all kinds of biobank research which were authorised by the concerned person (→ is this the donor/ research subject?).

Consequently, the existence of a general consensus to lower the level of protection of fundamental rights and freedom of patients/ research subjects – above all the right of self-determination – in order to foster the development of biomedical research could be inferred from this framework.

If the suspicion of a common tendency to move away from traditional research ethics can be confirmed – as it is argued by some scholars (Bernice and Caplan 2006) – then the priority issue concerning biobanking is discussing the normative validity of a new, emerging research ethics: biobanking research ethics.

From this perspective, the main question for biobanking research is the assessment of the normative legitimacy of consent forms (broad consent) allowing fewer human rights safeguards in favour of a greater expansion and exploitation of research.

The core of this issue is balancing research development and human rights protection. The question of how to balance these interests and the underlying principles is a meta-ethical issue: whether priority should be given to the interests of researchers – governed by the principle of beneficence – or to the interests of the individual (patient/ research subject) – governed by the principle of autonomy – is a question of outmost importance and difficulty that should be analysed and resolved on a meta-ethical level (Engelhardt 1996).

If the choice of an appropriate informed consent for biobanking research (both of retrospective and prospective research) and, more broadly, the definition of an appropriate ethical paradigm for biobank research really are a meta-ethical issue, then the common request of scientists, ethicists and policy makers to adopt a specific legislation for biobank turns into a secondary issue. This means that first of all, any form of regulation concerning research using biomaterials of human origin and/ or associated information should be legitimated and clearly defined on a meta-ethical level. This way the hierarchy of ethical principles could be applied to biobanking research.

After having decided what kind of ethical principle should govern biobanking research – whether the respect for individual autonomy or the safeguard of benefi-cence – the proposal of specific regulations for biobanking could be considered logically and function as an essential instrument to harmonise the existing mess of laws, declarations and ethical statements.

But first and foremost, it seems indispensable to clarify what kind of ethics is legitimate and should be applied to biobanking research.

References

Bernice, SE., Caplan, AL. (2006). Consent and anonymization in research involving biobanks. *EMBO Reports*, 7, 1-6

Council for International Organisations of Medical Sciences (2002). International Ethical Guide-lines for Biomedical Research Involving Human Subjects. www.cioms.ch. Accessed 12 June 2007

Council of Europe (2005). Additional Protocol to the Convention on Human Rights and Bio-medicine Concerning Biomedical Research. http://conventions.coe.int. Accessed 21 May 2007

Council of Europe (2006). Recommendation Rec (2006) 4 on research on biological materials of human origin. www.coe.int. Accessed 20 June 2007

Deschenes, M., Cardinal, G., Knoppers, BM., et al. (2001). Human Genetic Research, DNA banking and consent: a question of 'form'?. *Clin Genet* 59, 221-239

Engelhardt, IIT. (1996). *The Foundations of Bioethics*. New York: Oxford University Press

Hoeyer, K., Olofsson, BO., Mjorndal T et al. (2005). The Ethics of research using biobanks. Reason to question the importance attributed to informed consent,. *Arch Intern Med.* 165, 97-100

EMEA CHMP (2005). Concept paper on the development of a guideline on biobank issues rele-vant to pharmacogenetics. www.emea.eu.int. Accessed 23 April 2007

Knoppers, BM. (2005). Consent Revisited: Points to Consider. *Health Law Review* 2/3, 33-38

Loft, S., Poulsen, He. (1996). Cancer risk and oxidative DNA damage in man. *J Mol Med.* 74, 297-312

Maschke, KJ. (2005). Navigating an ethical patchwork – human gene banks. *Nature Biotechnol-ogy.* 5, 539-545

Merz, JF. (2003). On the intersection of privacy, consent, commerce and genetics research, in: BM Knoppers, ed., Populations and genetics: Legal Socio-Ethical Perspectives. New York: Kluwer Legal Int'l, 2003, 257-268

National Bioethics Advisory Commission. (1999). Research Involving Human Biological Mate-rials: Ethical Issues and Policy Guidance, in: Vol. I. Rockville, MD, USA

The Human Genome Organisation. (1998). The HUGO Ethics Committee, Statement on Dna Sampling in: www.hugo-international.org/. Accessed 11 September 2007

Thomas, HM. (2006). Key issues and questions in research with human biological materials. www.onlineethics.org. Accessed 1 September 1 2007

United Nations Educational, Scientific and Cultural Organization. (1997). The Universal Declaration on the Human Genome and Human Rights. http://portal.unesco.org/en/. Accessed 12 September 2007

World Health Organisation. (1998). Proposed International Guidelines on Ethical Issues in Medical Genetics and Genetic Services. http://whqlibdoc.who.int. Accessed 24 June 2007

Table 1: International, European, National Laws, Guidelines and Regulations on the use of human biomaterials and/or associated data for research purposes

ORGANISATIONS/ COUNTRIES	LAWS (L), GUIDELINES (G), REGULATIONS (R)	INFORMED CONSENT REQUIREMENTS
World Health Organisation	(G) Guideline for obtaining informed consent for the procurement and use of human tissues, cells, and fluids in research (2003)	Specific informed consent Partially restricted consent Broad consent
Council for International Organisations of Medical Sciences	(G) International ethical guidelines for biomedical research involving human subjects (2002)	Specific informed consent
United Nations Educational, Scientific and Cultural Organisation	(G) International declaration on human genetic data (2003)	Partially restricted consent
Human Genome Organisation	(G) Statement on DNA sampling: access and control (1998)	Broad consent
Council of Europe	(L) Convention for the protection of human rights and dignity of the human being with regard to the application of biology and medicine (1997) (G) Recommendation Rec (2006) 4 on research on biological materials of human origin (2006)	Specific informed consent
National Bioethics Advisory Commission	(G) Research involving human biological materials: ethical issues and policy guidance (1999)	Multi layered consent
Australia	(G) National statement on ethical conduct in human research (2007)	Specific informed consent Partially restricted consent Broad consent
Estonia	(L) Human genes research act	Broad consent
France	(G) Ethical issues raised by collections of biological materials and associated data: "Biobanks", "Biolibraries" (2003) – National consultative bioethics committee for health and life sciences (CCNE)	Specific informed consent
Germany	(G) Biobanks for research – National ethics council opinion (2004)	Broad consent
Italy	(G) Biobanks and research on human biological material – National Bioethics Committee Opinion (2006) (G) Guideline for clinical protocols of genetic research – Italian Society of Human Genetics (2006)	Partially restricted consent
	(G) Guideline for clinical protocols of genetic research – Italian Society of Human Genetics (2006)	Specific informed consent
	G) Guideline for genetic biobanks – Telethon	Specific informed consent
	(G) Guideline for the establishment and accreditation of biobanks	Specific informed consent
Japan	(G) Ethical guidelines for analytical research on the human genome/genes (2001)	Broad consent
Switzerland	(G) Biobanks: Obtainment, preservation and utilisation of human biological material	Broad consent Specific informed consent
Spain	(R) Royal decree 411/1996, by which activities regarding the use of human tissues are regulated (1996)	Informed express consent

United Kingdom	(L) Human tissue act (2004) (G) Human tissue and biological samples for use in research – Medical Research Council (2001)	Broad consent
Netherlands	(L) Civil code, Article 467 (1994) (G) Code for proper secondary use of human tissue in the Netherlands (2002)	Informed express consent
Iceland	(L) Act on biobanks No. 110 (2000)	Broad consent
Denmark	(L) Law on Biobanks No. 312 (2003)	Informed express consent
Sweden	(L) Law No. 297 (2005)	Specific informed consent
Norway	(L) Act on biobanks (2003)	Informed express consent

Box 1: Definition of informed consent models for biobanking research in light of the characterisation utilised in international literature and ethical documents

Informed consent model	Definition
Broad consent	This model of consent allows the use of biological specimens and related data in immediate research and in future investigations of any kind at any time in the future
Specific informed consent	This model of consent allows the use of biological specimens and related data only in immediate research. It forbids any future study that are not foreseen at the time of the original consent
Partially restricted consent	This model of consent allows the use of biological specimens and related data in specific immediate research and in future investigations directly or indirectly associated with them
Multi-layered consent	This model of consent requires to explain to a research subject several options in a detailed form

Once Given – Forever in a Biobank?

Legal Considerations Concerning the Protection of Donors and the Handling of Human Bodily Materials in Biobanks from a Swiss Perspective

Bianka S. Dörr

Abstract This article contains an outline of the current legal framework for research with biological material and personal data in Switzerland, with a focus on the preliminary draft proposal of a planned federal act on research involving humans regarding the rules to be applied to biobanks as well as the internationally acclaimed medico-ethical guidelines and recommendations of the Swiss Academy of Medical Sciences (SAMS) regarding biobanks.

In the Swiss legal system there are a number of different instruments to meet the challenge of protecting the rights of substance donors. In the context of biobanks, the debate concentrates on personality and property rights and the concept of informed consent. These instruments will be presented and discussed in this article.

As a new model, the establishment of an independent Biobank Ombuds Office is proposed to ensure that donors' rights are respected and the trust of donors and the public in biobanks is promoted.

1 Introduction

Whether you enter the term "biobank" in the query box of a search engine or check for publications on the topic in the last few years, you will find an overwhelming abundance of material.

This raises the question of what is so "special" about it. Why has the removal of human biological material, their storage in biobanks and use in research contexts become a subject so widely discussed and such a focus of political, social and scientific debate? After all, collections of samples of human blood, tissue, cells and organs gathered over centuries yet evidence a long established practice of collecting, storing, analysing and using any type of human biological material for research, education and other purposes (Zentrale Ethikkommission 2003, 2; SAMS, I.).

At least one answer is obvious: the successful sequencing of the human genome, the possibility of decoding human genetic material and the rapid progress in molecular genetic and biotechnological analytical processes. These developments together have led to a changed perception of the human body and increased the value of collections of biological material considerably. This is one explanation for the continuous expansion of existing collections and the promotion, on a national and international level, of new collections of biological materials, some of which strive for a substantial number of samples and genetic data (Mand 2005, 565; Dörr 2007, 449 ff). Moreover, the storage of human biological material in biobanks provides an indispensable source of biological research material for medical progress (Nationaler Ethikrat 2004, 36). In order to carry out significant medical and clinical studies and research complex illness mechanisms, scientists are increasingly dependent on using stored human biological material and data (Nationaler Ethikrat 2004, 32 ff). New is, however, the abundance and quality of data to be extracted from these collections and the possibility of collating and linking personal information with genetic information, especially by creating genetic and personality profiles, about the donors of biological material and their families to an extent previously unimaginable.

2 Problem Areas

The continuous expansion of small, medium and large sized biobanks expresses high hopes and visions for medical research and science. Despite all gain in knowledge that can be achieved therefrom, the "biobanking boom" often is a cause of unease and worry for a lot of people. The primary concern is the protection of the personality and privacy of substance donors, on the basis that a genetic analysis of the stored material reveals personal and very intimate data about the donors, and that research institutes today are increasingly linked to one another, exchanging samples and data on a national and international level. The fear is that biological material and the appertaining data may be used for a purpose donors have not expressly consented to, that the request for confidentiality of genetic data is not respected, that biological materials and data are used for commercial purposes or that any third party may access the data (Wellbrock 2003, 78; Nationaler Ethikrat 2004, 10 f; Antonow 2006, 54; SAMS, I.; Wicklein 2007, 5; Dörr 2007, 458). Moreover, in many cases it is not apparent for how long the biological material and personal data are intended to be stored in biobanks or for how long they are to be accessible to researchers – 5 years, 10 years or even a lifetime?

The set of problems outlined raises a number of interesting questions, but in light of the worldwide increase in biobanks, it calls primarily for a thorough examination of the legal scope and boundaries for research with human biological material and personal data, particularly regarding the use of stored material and data, the protection of donors' rights and the establishment and operation of biobanks. As a result of the lack of specific legislation and case law in this field, many scientific projects are confronted with considerable legal uncertainty when starting their collections and using biological material and data for research (Simon et al. 2006, 2; Wicklein 2007, 5). National and international legislators are therefore faced with the urgent challenge of establishing a coherent legal framework that balances the conflicting demands between the rights of donors and the interests of researchers and third parties.

In Switzerland, first steps have been taken: a preliminary draft proposal of the planned federal act on research involving humans as well as medico-ethical guidelines regarding biobanks have been elaborated in 2006. These measures will be briefly outlined in the second part of this contribution. Prior to that, however, I will discuss different legal instruments enshrined in the Swiss legal system in order to show how they meet the challenges of protecting the rights of donors of biological material. Following that, I propose the establishment of an independent Biobank Ombuds Office to ensure that donors' rights are respected and trust of donors and the public in biobanks is promoted.

3 Protection of Donors of Biological Material and Data …

Based on the awareness that donors' trust and their willingness to provide biological material and data is very much dependent on how well protected the samples and data will be from improper access and whether they are solely used for the agreed purpose, safeguarding the protection of the rights of the donors is an essential issue when establishing and operating a biobank. There are not merely biological materials of donors stored in biobanks, but also very sensitive personal data (about their genetic make-up, phenotype, diseases, lifestyle), which are either gained from the biological material itself or from the entry questionnaires. In particular, the genetic data may affect the personality of the substance donors notably and, due to their potential predictive nature, may also have a serious impact for their life plans. Moreover, it must be remembered that genetic data not only shed light on intimate and confidential information about the substance donors themselves but also always bear meaningful information with regard to genetically related individuals such as parents, siblings and (future) children (Schneider 2003, 2; Dörr 2007, 445).

In the Swiss legal system there are several instruments available to meet the challenge of protecting donors' rights. In the context of biobanks, the debate focuses mainly on personality rights and property rights as well as the concept of informed consent. Below, I will briefly trace and discuss this debate from the perspective of Swiss Civil Law and propose, as a new model for ensuring and strengthening trust, the establishment of an independent Biobank Ombuds Office.

... by personality rights and property rights

The dualism of the law of persons and the law of property is inherent in the Swiss system of private law. A 'person' is a legal subject vested with a human will and as such a carrier of rights and obligations (Riemer 2002, N 18; Schmid 2001, N 68 f, 550), while 'things' are defined as legal objects that have to serve this human will in one form or the other (Schmid 2001, N 73; Schmid and Hürlimann-Kaup 2003, N 4; Rey 2007, N 66; Büchler and Dörr 2008, 386).

Under Swiss law, the living human body as a whole is protected by the law of personality rights and is therefore not a thing (Rey 2007, N 101; Kälin 2002, 64; Riemer 2002, N 339). Thus, the creation of property rights in the living human body is precluded. However, as far as separated human biological materials are concerned, the prevailing opinion is that these qualify as mere things (Wiegand 2007, N 18; Rey 2007, N 106; Breitschmid 2003, 15 ff). With the separation from the human body, property in the bodily material transfers automatically to the person from whom it was separated; property originates along the lines of the substantial principle (*Substantialprinzip*) anchored in art. 643 Swiss Civil Code (Rey 2007, N 107). Although this approach may at first appear plausible as classified under property law, considering that separated bodily material becomes a thing, it is still unclear as to how the legal transformation from a person into a thing may occur without parallel loss of personality rights. A new approach in Switzerland (Büchler and Dörr 2008, 390 f) pleads for the continuity of personality rights derived from the link between the separated bodily material and the former carrier cumulatively to the classification of this material under property law. Links emanating from the fact that the extracted bodily substance formerly belonged to the human body as a whole and its personality and that genetic material can provide insight and information about this individual. In other words: the elements that cause protection under the law of personality are the information immanent to the genetic material as well as the possibility of identifying the carrier of the substance (Büchler and Dörr 2008, 391).

Thus, the possible dualism of property rights and personality rights means that, irrespective of the answer of who owns the bodily substance in each individual case, personality rights interests of the former substance carrier, which may be affected or violated by the concrete use or application, are to be protected. As far as personality rights continue to exist on separated bodily material, the use and further use of these materials by third parties is dependent on the consent of the former substance carrier (Büchler and Dörr 2008, 391).

... by informed consent

Personal and genetic data as well as human biological material stored in biobanks are subject to the protection of personality rights, in particular to the right of (informational) self-determination, and the pertinent data protection regulations.

Their storage and use in a research context demands the explicit and informed consent of the substance donors concerned, which needs to be voluntarily declared prior to any action and ideally given in writing (Haas 2007, 154 ff; Aebi-Müller 2005, 112 ff; Schmid 2001, N 873; Hausheer and Aebi-Müller 2008, 162 f).

In order to consent validly Swiss law further requires substance donors to be capable of judgment (*Urteilsfähigkeit*) (Bucher 1985, 43 ff; Guillod 1986, 39 ff; Haas 2007, 85 ff). Validity of consent also demands some degree of certainty (*Bestimmtheit*) as to the information given to substance donors (Haas 2007, 200 ff). For this reason, substance donors must, *inter alia*, have been informed in advance about the objectives of the sample collection, the type of use to which the biological material and personal data will be put, conditions and period of storage, measures for privacy and data protection, access by third parties, transfer of samples and data abroad as well as any intended commercial use. Should there still remain uncertainties at the time of consenting, any of these must be addressed transparently to the persons concerned.

As a basic principle and in the present context, a given consent is freely revocable at any time. This follows from the right to self-determination, which is substantiated by the consent: each individual must always be able to decide on the protection of one's own personality (Haas 2007, 185 ff; Rey 2007, N 766; Aebi-Müller 2005, 110; Schmid 2001, N 872). Withdrawing an informed consent is the only means by which donors obtain the possibility – even if they have generally agreed to the use of their biological materials and personal data – to exercise some form of control over the use of their data and materials.

In order to prevent the "jeopardy" of stalled research projects or established results being undermined by the possibility of withdrawing an informed consent at any time, it is recommended that the right to withdraw shall effect the future use of the biological material and personal data concerned as well as non-anonymised data. Where a withdrawal of consent does occur, it's negative effects might be limited by either permitting continued use for the running project on an anonymous basis or by destroying all existing biological material and personal data of a person but continuing work with those results and analyses already achieved prior to the withdrawal until the project is completed. However, it is always necessary to inform donors at the time of consenting about the fate of their biological material and personal data in the event of a withdrawal.

... by establishing a Biobank Ombuds Office

As described, Swiss law provides various instruments for the protection of donors of biological material (personality and property rights, informed consent), which have to be exercised by the donors themselves. In particular, donors rely that their rights are respected by researchers and trust that the provided material will be used according to the stated purpose. However, whether this is really the case, is extremely difficult to control for them. Therefore, I propose that a Biobank Ombuds

Office be created as an independent forum for advice, control, complaints and mediation, as a self-regulating means of safeguarding, ensuring and strengthening public and donors trust in and promoting transparency of biobanks. This Biobank Ombuds Office could be equipped with diverse competences but its main tasks would consist in safeguarding the protection of the rights of donors of biological materials. Precisely, a Biobank Ombuds Office could be entrusted with the following mission and responsibilities: ideally, it would already play an advisory and accompanying role in the design phase of research projects and provide expertise in assessments and competitions for research grants. This way, the Biobank Ombuds Office could, at an early stage of project planning, safeguard that their rights are observed. Additionally, the Biobank Ombuds Office would provide help to those donors who feel that their rights have been violated, inform them about the relevant legal principles and discuss next steps or otherwise advise on a specific legal situation. Besides, it would seem reasonable to entrust such a Biobank Ombuds Office with the task of monitoring biobanks to ensure the fulfilment of requirements regarding the protection of personality rights and data. Such an Office could also provide valuable services in connection with (specified purpose) consents or their withdrawal, especially by ensuring that data have been anonymised or completely destroyed.

In conflicts, the Biobank Ombuds Office would play a mediating role, always endeavouring to reach an amicable and fair settlement between the parties. As for the general public, this Biobank Ombuds Office would provide information and answer questions on relevant issues linked to biobanks. Yet another important competence that should be given to the Biobank Ombuds Office, is an unlimited right to information and inspection as well as, in case of necessity, the right to claim access to files or carry out site visits. The Biobank Ombuds Office would follow up on and denounce deviations from the "Good Laboratory Practice" uncovered by its own initiative or through complaints, and generally keep in contact with biobanks to check on the progress of research projects.

Moreover, the Biobank Ombuds Office should be in a position to give recommendations. Such recommendations would have moral authority, but should be made reasonably accessible to the public in some form, for example as an annual report or a newsletter. The Biobank Ombuds Office would not, however, have any legal authority to enforce sanctions. Finally, the services of the Biobank Ombuds Office should be offered free of charge for donors of biological material and data.

4 Current Legal Framework in Switzerland[1]

As mentioned at the outset, the steady growth in the number of biobanks and consequent ethical, legal and social challenges in turn increases the need for clarification of the legal foundations regarding research with human biological material and personal data, the (further) use of already collected biological material as well as the set-up and operation of biobanks. In Switzerland, as in a number of other European countries, there is currently no uniform legislation or regulation of research involving human biological material and personal data at federal level nor a pertinent legal practice.

Therefore, many scientific projects are faced with open legal issues when starting collecting and using biological samples and data in research. Legislators are thus called upon to meet these challenges and create a coherent legal framework for research with human biological material and personal data in biobanks. In Switzerland, efforts are being made to remedy this situation; the legislator is currently working on a comprehensive act on research involving humans (*Humanforschungsgesetz*) as well as on a corresponding constitutional provision. Until these efforts result in the passage of the act on research involving humans, the general rules and legal principles of the existing cantonal, federal and constitutional laws, in particular personality and data protection laws, apply.

4.1 In preparation: Federal Act on Research involving Humans

Preliminary Draft Proposal of a Federal Act on Research involving Humans (Vorentwurf Humanforschungsgesetz)

In February 2006, the Federal Council opened consultation on the preliminary draft proposal of a planned federal act on research involving humans as well as a new corresponding constitutional provision. The latter intends to grant the Confederation comprehensive competence to regulate in this area. The planned act itself will substantiate the new constitutional provision and aims for a uniform and extensive regulation on federal level in the field of research involving humans.

The preliminary draft proposal uses a broad concept of "research involving humans"; this means that the research scope includes not only research with humans but also extends to biological material of human origin and personal data (Erläuternder Bericht zum Vorentwurf 2006, 75 f). The relevant provisions are to be found in chapter 4, chiefly in articles 35–49 of the preliminary draft proposal. Covered are all research-related activities with biological material and personal data: from the removal of material or the collection of personal data (articles 38–

---

[1] A detailed analysis of the Swiss legal framework regarding biobanks can also be found in the article Dörr 2007, 466 ff (in German language).

40) to its use for research purposes (articles 41–46) as well as its storage in and the operation of biobanks (articles 47–49). These articles are generally applicable to situations in which biological material and personal data have already been collected and are to be used for further research purposes; however, the provisions on research involving humans (chapter 2 and 3) are applicable if biological material and personal data are to be removed and collected as part of a research project (Erläuternder Bericht zum Vorentwurf 2006, 96).

The regulations in detail

The detailed regulations on research with biological material of human origin and personal data (articles 35–49) reflect the efforts of the legislator to reconcile the protection of the personality, bodily integrity and privacy of the donors with the freedom of scientific research and the significance of research for human health and society as a whole. The area of conflict between the protection of the personality and the freedom of scientific research is particularly noticeable in the context of biobanks and therefore requires careful balancing.

The first paragraph of the regulations in the chapter on research with biological material and personal data comprises general provisions such as the principle of subsidiarity (art. 35), the right to know and the right not to know (art. 36) as well as regulations on the foreign export of biological material and personal data (art. 37). Subsidiarity as understood in art. 35 means that carrying out a research project with uncoded biological material and personal data is only allowed if equivalent results cannot be achieved with anonymised material and data. Moreover, research with uncoded biological material and data requires a special justification, because of the imminent risk that the privacy of the persons concerned is violated (Erläuternder Bericht zum Vorentwurf 2006, 96). The right to know and the right not to know as anchored in art. 36 are expressions of the right of self-determination; these rights enable the persons concerned to either take notice of information concerning their health or to ignore them. Art. 37, which permits the foreign export of biological material and personal data for research purposes only in an anonymised or coded fashion and provided equivalent regulations apply to the use or further use of the material in the country of destination, serves the purpose of protecting the personality, privacy and personal data of substance donors.

Art. 38 rules that the removal of biological material or the collection of personal data requires sufficient information and the written consent of the persons concerned prior to the beginning of the research project; these people must be given adequate time for consideration before consenting. According to art. 39, the person concerned can withdraw his or her consent informally at any given time and without reasons as well as forbid the use of his or her coded or uncoded biological material and personal data (Erläuternder Bericht zum Vorentwurf 2006, 98). In this event, the biological material or the personal data may, in anonymised form, be further used for the running project.

As to the further use of biological material and personal data for research purposes, the substance donors are required to have given consent, which, depending on the degree of anonymisation of the material and personal data, may be in the form of an objection (art. 41), a general consent (art. 42) or a specific consent (art. 43). Are biological materials or personal data be further used in coded or uncoded fashion, the person concerned has the right to refuse or withdraw his or her consent (art. 44). In cases in which consent cannot be obtained for the further use of human biological material or data, research may be allowed without consent if the project pursues an essential purpose (lit. a), if it is impossible or unreasonable to get the consent or inform about the right to withdrawal (lit. b) or, if there exists no documented refusal (lit. c) (art. 45). The circulation of material and data for other than research purposes is laid down in art. 46.

In the last section of chapter 4 (articles 47–49) the preliminary draft proposal foresees specific regulations regarding the operation and organisation of biobanks. The applicable principle, in the interest of the quality of the research, is that the operator of a biobank in order to undertake this activity must ensure that he or she fulfils the necessary technical and operational requirements (art. 48, par. 1). Additionally, he or she has to state in writing (art. 48, par. 2): a) the purpose of the biobank; b) the criteria for collection, use of material and length of storage; c) the organisation and distribution of responsibilities, in particular how supervision and control of the biobank are to be regulated and how quality is to be guaranteed; d) the conditions of a transfer of samples to third parties, this even before the biobank starts its operation and e) how data protection is to be guaranteed (Erläuternder Bericht zum Vorentwurf 2006, 103 ff). Further, the operation of a large-scale biobank (art. 49) requires an official operating policy that is accessible to the public and subject to authorization (art. 57).

Results of the consultation procedure

In the course of the four-month consultation procedure that ended May 31st 2006, the preliminary draft proposal on research involving humans raised a variety of reactions among the participants. Out of a total of 153 statements submitted by the participating parties, two-third of them evaluated the preliminary draft proposal positively or neutrally, while one-third gave it a negative appraisal (Bericht Vernehmlassungsverfahren 2007, 11 ff). In particular, the regulations on research with biological material and personal data raised some very critical comments: Among the suggestions expressed were those who demanded a complete revision of the corresponding provisions in accordance with the biobank guidelines of the Swiss Academy of Medical Sciences (see 4.2) or proposed a separate regulation of research with biological material and personal data (Bericht Vernehmlassungsverfahren 2007, 32).

Law-making process

Following the normal run of the law-making process the preliminary draft proposal on research involving humans will undergo an intense revision next. At this stage, it is impossible to predict how and to what extent requests for modification or restructuring that were articulated during the consultation procedure will be integrated into the draft bill expected for late 2009. It is assumed, however, that the present text of the preliminary draft proposal will experience substantial change, especially with regard to the regulations applicable to biobanks, before the final version is adopted. The entry into force of the federal act on research involving humans will most probably not take place before the year 2013.

4.2 Canons of Professional Ethics

General information

In May 2006, the Swiss Academy of Medical Sciences (SAMS) published the medico-ethical guidelines and recommendations *Biobanks: obtainment, preservation and use of human biological material* to provide guidance and orientation until the planned federal act on research involving humans is elaborated and ready for becoming effective (SAMS, I.). As regards content, the guidelines contain provisions to protect the substance donors' dignity and personality as well as regulations to guarantee the quality and safety of biobanks. These guidelines have attracted great interest both on a national as well as an international level and are cherished for balancing donors and researchers interests well.

For many years now, the SAMS, founded in 1943, has established guidelines and recommendations on topical medical issues and has clarified ethical questions related to contemporary biomedical and technical achievements and their social consequences. The SAMS guidelines and recommendations are valued for always being up-to-date and for their flexibility, which allows for quick reactions to social changes (Rüetschi 2004, 1222). However, they are, strictly speaking, not legally binding, but have nonetheless acquired indirectly certain importance: several cantons refer to them in their cantonal health regulations – usually via cross reference – and courts regularly use them as orientation guide (Schwander 2005, 58; Rüetschi 2004, 1223; Schöning 1995: 36 f, 57 ff).

Addressees of the biobank guidelines and area of application

The SAMS biobank guidelines are addressed to all operators and users of public and private biobanks, as well as other collections of human biological material. They apply to research, teaching, education, and training (SAMS, II.2). However, they are not applicable to the use of tissue material for individual diagnosis, therapeutic or forensic purposes, or quality control and safety, as long as this occurs within the medical practice (SAMS, II.2). The content of the guidelines is based on the premise that transparent rules strengthen trust in research in the long run and increase the willingness of potential donors to donate their biological material (Dittmann and Salathé 2006, 1023).

The regulations in detail

According to the preamble, the biobank guidelines aim to consider both the present national and international framework and the essential principles of bioethics (autonomy, welfare, justice). In particular, the guidelines are clearly designed to balance individual rights versus research interests. Thus, they contain provisions for the protection of dignity and personality of donors as well as for quality control in the operation of biobanks. The guidelines prescribe, for example, that donors of biological material must consent to the extraction, storage and use of their samples for research purposes and that this (written) consent must be preceded by a written information (SAMS, II.4.2, 4.3). The extent of this information has, however, to be proportionate to the intended use of the samples and data (SAMS, II. 4.2); a corresponding list of reference points that are considered particularly relevant and important for donors is contained in section 4.2 of the guidelines. As a rule, consent is given in form of a general consent, allowing the further use of samples and data for future research projects under certain conditions (SAMS, II. 4.3). However, a right to withdraw this consent at any given time is foreseen for non-irreversibly anonymised samples as well as for the future use of samples and data; in the event of a withdrawal, the samples have to be destroyed, not the obtained results (SAMS, II. 4.6). Moreover, the biobank guidelines provide donors with the right to be informed of any diagnostically and therapeutically relevant results (right to know) or to ignore, partially or fully, this information (right not to know) (SAMS, 4.7).

All research projects that may directly affect donors of biological material require the formal approval of the relevant ethics commission for clinical trials (SAMS, 4.1). In addition, biobanks need to have an appropriate quality control and safety system to ensure that stored material and data are preserved and managed faultlessly under observance of the rights of the substance donors and in accordance with the principles of Good Laboratory Practice (SAMS, II.3, 3.1, 3.2).

The personnel must be adequately qualified and the appropriate infrastructure available (SAMS, II.3). The essential aspects, such as organisation, responsibilities, area of application for the biobank, purpose, origin of the stored samples, people with access rights, have to be defined in rules of procedure (SAMS, II.3, 3.4).

As a conclusion, the SAMS guidelines address the following recommendations: 1) Creation of registers of public and private biobanks; 2) Establishment of standards for training in laboratory work; 3) Establishment of terms for the accreditation of biobanks; and 4) Creation of an information and consent form for patients entering a hospital (SAMS, IV.)

5 Conclusion

Biobanks are considered an invaluable source of information for biomedical research and a potential that is still to be exhausted for progress in genetic epidemiology and for the development of future therapeutic and diagnostic applications in medicine. The establishment of biobanks is associated with numerous ethical, legal and social challenges and the need for clarification is constantly increasing with regard to research involving human biological material and personal data, the handling and use of stored biological materials, the protection of donors and the building up and operation of biobanks. National and international legislators are therefore called upon to face these challenges with a coherent legal framework.

One way of strengthening and ensuring trust as well as the protection of substance donors could be the establishment of a Biobank Ombuds Office, as an independent forum for advice, control and complaints. Ideally, this Biobank Ombuds Office would play a mediatory role between substance donors, researchers, authorities and the general public and provide primary support to donors in exercising their rights.

The Swiss authorities have recognized the need to take action in the context of biobanks and have made first efforts by elaborating draft laws and canons of professional ethics with a view to meeting the challenges raised. As early as in 2006, the first preliminary draft proposal of a planned federal act on research involving humans and the medico-ethical guidelines and recommendations on biobanking by the Swiss Academy of Medical Sciences were presented. It would be desirable, if the Swiss regulations and recommendations on biobanks as well as the idea of establishing a Biobank Ombuds Office will promote new motivation and inspiration for further debate on the topic of in Germany and other European countries.

References

Aebi-Müller RE (2005) Personenbezogene Informationen im Schutz des zivilrechtlichen Persönlichkeitsschutzes. Unter besonderer Berücksichtigung der Rechtslage in der Schweiz und in Deutschland. Stämpfli, Bern

Antonow K (2006) Der rechtliche Rahmen der Zulässigkeit für Biobanken zu Forschungszwecken. Nomos, Baden-Baden

Bericht über die Ergebnisse des Vernehmlassungsverfahrens zum Vorentwurf einer Verfassungsbestimmung und eines Bundesgesetzes über die Forschung am Menschen (2007). Bern

Breitschmid P (2003) Wenn Organe Sachen wären … Einige unorthodoxe Überlegungen zu einer noch zu diskutierenden Frage. In: Honsell H et al (eds), Aktuelle Aspekte des Schuld- und Sachenrechts. Festschrift für Heinz Rey zum 60. Geburtstag. Schulthess, Zürich

Bucher E (1985) Der Persönlichkeitsschutz beim ärztlichen Handeln. In: Wiegand W (ed), Arzt und Recht. Stämpfli, Bern

Büchler A, Dörr BS (2008) Medizinische Forschung an und mit menschlichen Körpersubstanzen. Verfügungsrechte über den Körper im Spannungsfeld von Persönlichkeitsrechten und Forschungsinteressen, ZSR 4: 381-406

Dittmann V, Salathé M (2006) Übergangsregelung für Biobanken bis zum Inkrafttreten des neuen Humanforschungsgesetzes, SAEZ 23: 1022-1023

Dörr BS (2007) Blut, Gewebe, Zellen, DNA und Daten – Biobanken im Spannungsfeld von Persönlichkeitsrechten und Forschungsvisionen. In: Dörr BS, Michel M (eds), Biomedizinrecht. Entwicklungen Perspektiven Herausforderungen. Dike, Zürich

Erläuternder Bericht zum Vorentwurf eines Bundesgesetzes über die Forschung am Menschen (2006). Bern.

Guillod O (1986) Le consentement éclairé du patient – autodétermination ou paternalisme? Editions Ides et Calendes, Neuchâtel

Haas R (2007) Die Einwilligung in eine Persönlichkeitsverletzung nach Art. 28 Abs. 2 ZGB. Schulthess, Zürich

Hausheer H, Aebi-Müller RE (2008) Das Personenrecht des Schweizerischen Zivilgesetzbuches, 3rd ed. Stämpfli, Bern

Kälin O (2002) Der Sachbegriff im schweizerischen ZGB. Schulthess, Zürich

Mand E (2005) Biobanken für die Forschung und informationelle Selbstbestimmung, MedR 10: 565-575

Nationaler Ethikrat (2004) Biobanken für die Forschung. Berlin

Rey H (2007) Die Grundlagen des Sachenrechts und das Eigentum, 3rd ed. Stämpfli, Bern

Riemer HM (2002) Personenrecht des ZGB: Studienbuch und Bundesgerichtspraxis, 2nd ed. Stämpfli, Bern

Rüetschi D (2004) Die Medizinisch ethischen Richtlinien der SAMW aus juristischer Sicht, SAEZ 23: 1222-1225

SAMS (Schweizerische Akademie der Medizinischen Wissenschaften/Swiss Academy of Medical Sciences) (2006) Biobanken: Gewinnung, Aufbewahrung und Nutzung von menschlichem biologischem Material. Medizinisch-ethische Richtlinien und Empfehlungen. Basel

Schmid J (2001) Einleitungsartikel des ZGB und Personenrecht. Schulthess, Zürich

Schmid J, Hürlimann-Kaup B (2003) Sachenrecht, 2nd ed. Schulthess, Zürich

Schneider I (2003) Biobanken im Spannungsfeld zwischen Gemeinwohl und partikularen Interessen. Friedrich-Ebert-Stiftung, Berlin. Im Internet abrufbar unter http://fesportal.fes.de/pls/portal30/docs/FOLDER/STABSABTEILUNG/BIOBANKENSCHNEIDER.PDF

Schöning R (1996) Rechtliche Aspekte der Organtransplantation, unter besonderer Berücksichtigung des Strafrechts. Schulthess, Zürich

Schwander V (2005) Medizinische Forschung am Menschen zwischen Wissenschaftsfreiheit und Persönlichkeitsschutz. Zur Rechtslage in der Schweiz. In: Brudermüller G et al (eds), Forschung am Menschen: Ethische Grenzen medizinischer Machbarkeit. Königshausen & Neumann, Würzburg

Simon JW et al (2006) Biomaterialbanken – Rechtliche Rahmenbedingungen. Medizinisch Wissenschaftliche Verlagsgesellschaft, Berlin

Wellbrock R (2003) Datenschutzrechtliche Aspekte des Aufbaus von Biobanken für Forschungszwecke, MedR 2: 77-82

Wicklein M (2007) Biobanken zwischen Wissenschaftsfreiheit, Eigentumsrecht und Persönlichkeitsschutz. Tectum, Marburg

Wiegand W (2007) Vor Art. 641 ff. ZGB. In: Honsell H et al (eds), Basler Kommentar zum Schweizerischen Privatrecht, Zivilgesetzbuch II, 3rd ed. Helbing & Lichtenhahn, Basel

Zentrale Ethikkommission der Bundesärztekammer (2003) Stellungnahme: Die (Weiter-)Verwendung von menschlichen Körpermaterialien für Zwecke medizinischer Forschung. Berlin

Biobanks and the Law

Thoughts on the Protection of Self-Determination with Regards to France and Germany

Kathrin Nitschmann

Abstract The implementation of regulatory frameworks to protect the fundamental right to self-determination requires an interdisciplinary and international discourse about competing interests. Within this discourse, various national perspectives on the position of the individual come into play. The article looks at the French legal system, outlining the position of the individual vis à vis the handling of biobanks, with a focus on the right to self-determination. Some national particularities will also be described, as a means of providing background to current legal conceptions of the individual right.

1 Introduction

Examining and comparing foreign legal systems has become the inevitable task of jurists, especially in fields in which human activity may have serious implications for populations worldwide and where harmonization of regulation is required. Understanding the similarities and differences between legal systems is a fundamental condition of a successful process of harmonization. The act of comparing law systems reveals not only legal specialities but also differences in terminology: it was the discussion on the terminology of "biobanks" – in French "biobanque" – that first of all caught my eye. Considering the social power of language and communication in general and the images that terminology may transport, it is worth back-pedalling here briefly to address the question of terminology.

During a debate of the French senate it was pointed out that, despite the difficulties and differences of national systems, the opinion of the French National Ethic Committee (CCNE) on biobanks would be the first document of its kind to refer to the problem of translation (Comité consultatif national d'éthique 2003, 14).[1] In the opinion an "extremely illuminating debate" concerning terminology is mentioned, which results in the theory that the term "biolibrary" would be preferable to the term "biobank". The statement of the CCNE took into account the im-

---

[1] See also the joint document of Comité consultatif national d'éthique /Nationaler Ethikrat.

age "of deposit of property with a market value" associated with the idea of a "bank" and which "eclipses the human origins of the samples and the ethical problems which ensue", whereas the term "biolibrary" would emphasize the idea of archiving for common purposes.

However, it is not only the German equivalent "Biothek" that may in this context incite reflection on semantics: going through a report of the French "Assemblée Nationale" and the Senate I observed a virtually inflationary use of the prefixoid "bio" which was included in at least 30 compounds such as "bioterrorism", "biosecurity", "biovigilance", "bioinformatiques" (Assemblée nationale and Sénat 2005).

Anyway, aside from these comments on linguistics, the focus is on some legal aspects related to biobanks in the French context.

The mission of the law and legal intervention in this field is well known from other biomedical matters such as experimentation on human beings, stem-cell research, cloning and transplantation. In the centre of the debate is the individual, whose interests collide with research interests or public interests such as health and security.

In line with the "traditional" discussion of law and medical ethics, the first step is to sketch out the recurring question of the relationship between ethics, law and human genetics. It is imperative to determine, in a rational discourse, the extent to which the legislator should intervene in principles of fairness and morality, taking into consideration the effectiveness and the credibility of the law, especially where the ultima ratio of criminal law is concerned (Mason et al 2002).

Linked to these reflections is the question as to what extent interests of individuals or groups should be considered in the legislative process. The participatory model of the French "États Généraux", which will be briefly examined in step two, is in the forefront of the realization of subjective rights such as the right to self-determination. The relationship of the individual to the concrete right of self-determination in France, which is the focus of this paper, continues with a short description of some French national particularities regarding the role of the individual. This leads us to examine statutory law regarding the individual and the human body, expressly referred to in French legislation for the first time in 1994, which is relevant when discussing self-determination in the context of biobanks. It will be shown how the conflict between individual and public interests influencing the dimension of self-determination is manifested, particularly in the conception of consent as applied in French, as opposed to German criminal law. Finally, we will look at how the right to self-determination is handled within the regulations relevant to biobanks.

2 Ethics, Law and Human Genetics

The intervention of law in the field of medical practice and research is increasing in response to the extension of this field. It is the antagonism between various interests that implicates the involvement of the law (Mason et al 2002: 4): human rights vs. freedom of scientific research, medical paternalism vs. individual freedom and self-determination respectively. Against this background of interests it is impossible to detach the legal debate from the moral rules; they stand in a permanent and complex inter-relation with each other.[2]

The prioritisation of interests may vary in different legal cultures, which re-emphasises the value of the examination and comparison of legal systems as a whole.[3] So we have to keep in mind that the right to self-determination, for example, has a much higher rank in the Anglo-Saxon tradition than in that of continental Europe, where its conception is not homogenous, as we will see with regards to France and Germany. For decades, the right to self-determination has been a topic in bioethical discussions, re-arising regularly at the centre of the question as to whether it is guaranteed as a fundamental right in tension with various other constitutionally protected values or interests.

A specific characteristic of biobanks that implies a new challenge for the conceptualisation of the right to self-determination is their dimension in respect to content and time. Collections of biological materials hold, apart from an unimaginable fundus for scientific research, an immense economic capital as well as capital for public purposes such as public health, public security and justice. The significance of these interests has grown in step with genetic progress and seem to promote the development of regulatory mechanisms for the protection of the individual.

The appropriate balancing of interests underlies the principles of rationality and proportionality and must be conducted on both prospective and international bases in order to prevent the development of 'oasis research', generally characterised by a low standard of human rights protection. A major initiative in this regard is the common French-German statement of Ethics Committees.

In regard to legislative intervention in medical ethics, a certain restraint is recommended (Mason et al 2002: 21-22). A lesson can be learned here from the French legislation on embryonic stem cells and cloning,[4] which has been cast into doubt by the recent tendency towards a more liberal attitude in research matters. In its attempts to follow scientific exigencies, the 2004 French legislation on bioethics is considered to have disqualified itself, as far as the experimentation with

---

[2] See in general for the relation between law and ethics Ellscheid 1998; for the medical field Schroth 2004, Eser A 1988, Ellscheid 1979; for the French context Nitschmann 2007: 75-81.

[3] For the idea of "Wertende Rechtsvergleichung" see Jung 2005.

[4] Loi n° 2004-800 relative à la bioéthique, 06/08/2004, Journal Officiel 07/08/2004, 14040 ; a critical analysis of the development in this context give Bellivier et al 2006: 277.

embryos is concerned.[5] Law, as a reflection of the views of society, obviously adapts slowly to scientific progress, and legal authorities should avoid, if possible, the anticipation or domination of the process of social debate. This alludes to the topic of individual and collective participation in processes of legislation as a contribution to the realisation of individual rights. The French model of the "États généraux" is an example of participation of citizens at the lowest democratic level and is worth a brief closer look.

3 Participation of the Individual in Legal Debates

The idea of the French "États Généraux" is that the law embodies the public conscience and expresses a social consensus on principles and rights. The extent to which different social groups should be represented in the legislative process of a democratic constitutional state is a controversial issue, but, arguably, the gap between legislating authorities and the population can be diminished and the acceptance and sustainability of legal rules increased by analysing the object of regulation from this perspective.[6]

The objective of the États Généraux that originated at the beginning of the 14th century is to offer a platform for a public debate, in times of political crisis, concerning a certain subject.[7] This enables delicate questions with a special social significance to be discussed by citizens themselves and solutions or recommendations to be worked out according to their needs.

Whereas this form of public assembly was the exception in the past, it has become a common means of approaching social topics such as research and education, quality of the health system, culture etc. in French politics today and –in some cases – a means to enhance a legislative process. In regard to the preparation of legislation, the "États généraux" requires an enormous investment of time and financial as well as personal resources. When the "États généraux" prepared a law concerning the quality of the health system and patients' rights in 2002,[8] the debates lasted 7 months and were accompanied inter alia by a large information

---

[5] The French legislation on bioethics began in 1994 with an absolute interdiction of experimentation on embryos. The legislation of 2004 (art. L 2151-5 Code de la santé publique), with a decree of 2006, permits experimentation for a period of 5 years under certain conditions, which are said to be vague and therefore worthless as a regulation of research practices. For more details see Bellivier et al 2006: 277-288.

[6] For legislative processes in Germany see Schulze-Fielitz 1988: 255.

[7] For more details see Lefebvre G et Terroine 1953. The king of France called for these meetings – only 21 in 487 were counted – to discuss controversial political questions with representatives of the nation belonging to different social classes (Noblesse, Clergé, Tiers d'État). See also for the description of the "États généraux de la santé" Nitschmann 2007: 89-94: 320.

[8] Loi n° 2002-303 du 4 mars 2002 relative aux droit des malades et à la qualité du système de santé.

campaign, an internet platform and a central orientation committee (Nitschmann 2007a). This modus operandi suited the governing ideology of a "démocratie sanitaire", the ambition to create a health system offering the potential for participation, which was one of the figureheads of this legislation.[9]

Arguably, the need for participation on a lower, local level is heightened in a traditionally centralized state such as France and might collide with the principles of a representative democracy. Nevertheless, broader public information and organized public discussions on sensitive matters such as biobanks might result in an improved process for reaching a convincing legal consensus. The "États Généraux" might serve as a model that could be adapted with respect to the status of the individual and the rights that were referred to earlier.

4 The Unauthorized Individual – a Retrospective

To understand the right to self-determination in France, a retrospective outline of the national context is indispensable as background to the current conception of the right in the context of biobanks.

In France, the relationship between state and individual is suffused with the idea of an omnipotent 'paternal state, which is manifested by the pronounced supremacy of the central state over the totally subordinated citizen (Bonnin 1812: 4). This must be taken into account in the field of medical treatment, as the individual has the status of a "user of public service" and public law is applied in relation to professionals.[10] The relation between state authorities and citizens in France has an impressive history. For a long time, the citizen was regarded as completely minor and even designated an "invalid administrated", not able to act rationally and tending to disturb the "ordre public" (Foulquier 2003: 35-38). A group of objective impersonal rights and obligations concretized the citizen's position, but there was no guaranteed protection on the basis of subjective rights. Personal interests aside, the concepts of individual liberties and public interests were opposed to each other (Chevallier 1979: 3). The idea of subjective rights has developed slowly but was recently reinforced by the 2002 law on patients' rights and the health system.[11]

This evolution toward implementation of subjective rights is accentuated by the backdrop of pronounced paternalism in the relations between doctors and patients that seems to correspond to the traditional relation between the state and the individual.

---

[9] For the discussion on this terminology see Nitschmann 2007: 121-125.

[10] For the legal status of the individual in public hospitals see Dupont et al marginal numbers 457-464.

[11] The emancipation of the citizens illustrates Foulquier 2003 : 42; see loi n° 2002-303 relative aux droits des malades et à la qualité du système de santé, 04/03/2002, Journal Officiel 05/03/2002.

The picture of the patient drawn by the well-known medical practitioner Louis Portes, mid-20th century president of the Medical Association, was remarkable: he was described as blind and intellectually passive, "like a child to be tamed, certainly not to be cheated – a child to be consoled, not to be abused – a child to save, or simply to be healed."[12] The idea of self-determination was completely alien to this conception of the relation between doctor and patient; the latter was unable to provide consent.[13] Further, Portes assumed that not even a dialogue would be possible between the two protagonists (Portes 1954: 160).

The inferior position of the individual in these two examples has for a long time had its counterpart in a relatively weakly developed position in written law.

5 The Rise of the Individual in French legislation – the Statutory Acknowledgement of the Status of the Human Body

The status of the human body was affirmed for the first time in French law in the 1994 legislation on bioethics.[14] Under the headline "Respect of the human body", articles 16 to 16-13 declare the primacy of the person (art. 16) but decline the right of the person to the body. This article was introduced to counterbalance the idea that a human being may be a purely biological source of organs, cells or products, by emphasizing that it is an entity of body and person (Thouvenin 2001: 113-115). The elements of the person, as a part of the body, are included in the protection of human dignity guaranteed in art. 16 of the Civil Code.

Art. 16-1-2 affirms the inviolable nature of the human body, protected primarily in the criminal code.[15] This principle is complemented by the assertion, in art. 16-3, that a violation of bodily integrity is permissible only when medically necessary and in the interests of the person concerned, or with the prior consent of the violated individual, for the benefit of a third person.[16] The principle of inviolability is restricted in a situation in which the right to self-determination is reinforced.

---

[12] Portes 1954: 163; for an overview on the vision of Portes' see Nitschmann 2007: 60-72.

[13] Portes' negation of the possibility of consent was based on the idea that the disease would prevent the patient from taking a lucid decision, Portes 1954: 159-172.

[14] Loi n° 94-653 relative au respect du corps humain, 29/07/1994, Journal Officiel 30/07/1994, 11056 and loi n° 94-654, relative au don et à l'utilisation des éléments et produits du corps humain, à l'assistance médicale à la procréation et au diagnostic prénatal, 29/071994, Journal Officiel 30/07/1994, 11060.

[15] Art. 221-1 f., art. 222-1 f., art. 222-7 f., art. 222-22 f., 223-1 f.

[16] Art. 16-3 Code civil : Il ne peut être porté atteinte à l'intégrité du corps humain qu'en cas de nécessité médicale pour la personne ou à titre exceptionnel dans l'intérêt thérapeutique d'autrui. Le consentement de l'intéressé doit être recueilli préalablement hors le cas où son état rend nécessaire une intervention thérapeutique à laquelle il n'est pas à même de consentir.

The idea of a partnership between medical practitioners and patients regarding treatment decisions had been conveyed in the 1994 legislation and was confirmed in 2002. The law on the quality of the health system and patients' rights[17] introduced, in the public health code, a remarkable description of the relationship between doctor and patient (art. L. 1111-1 to 1111-8, especially L. 1111-4 al. 1 CSP) that stressed the autonomy of the individual during medical treatment and expressed a renunciation of the traditional paternalism that had dominated therapeutic situations for years.

The protection of self-determination also applies in the field of personal data. Information about an element or product of the human body relating to a donor is subject to the obligation of secrecy (art. 16-8 civil code). Further, since 2004 the examination of genetic characteristics is subject to consent (art. 16-10 civil code). The identification of a person on the basis of genetic characteristics is permitted in connection with criminal procedures and – again respecting the need for written consent – for medical or research purposes (art. 16-11 civil code). Discrimination on grounds of genetic characteristics is prohibited (art. 16-13 civil code).

However, tendencies to soften the right to self-determination can be observed in regard to genetic information. In 2004, the legislator invented a procedure for the transmission of medical information concerning familial gene characteristics (art. 1131-1 code of public health). Should a genetic anomaly be revealed, the person concerned may, after obtaining advice regarding the risks to family members, initiate a formal process for informing them. Legal responsibility for a failure to transmit information is excluded by the legislation. Credit for the current legal solution goes to the National Ethics Committee and others who argued explicitly against a proposal for genetic screening including an obligation to inform family members (Comité consultatif national d'éthique 2003).

The right to self-determination confronts another limitation when the handling of the body comes into contact with commercial purposes that endanger the integrity and, in particular, the dignity of a person. Accordingly, conventions are null and void if they attribute an economic value to a human body or its elements or require remuneration in relation to experiments on humans or the removal of body elements (art. 16-5, 16-6 civil code).

The regulations relating to the status of the person in 16 to 16-13 of the civil code are declared part of the "ordre public" (art. 16-9 civil code). The person is therefore protected as a part of the public order, on the one hand, but, on the other hand, protection may be subject to limitations imposed by other public concerns or interests of constitutional value, such as research, public health or security, come into play. The tension between public and individual interests is apparent in the concept of consent in criminal law.

---

[17] The idea of the participatory model described by Katzenmeier 2002: 57-61; regarding the legislation of 2002 see Nitschmann 2007: 183-185.

6 Self Determination vs. Public Order – the Concept of Consent in French Criminal Law

To understand the difference between French and German perceptions of the right to self-determination, it is useful to have a closer look at its implementation, particularly in regard to the notion of "consent" in the criminal law.

The French historical conception of consent is based on a traditional social perception that a violation of the individual implies a violation of society as an entity (Garraud 1828: 130). The right to dispose of one's body cannot be considered as self-evident in the history of the law (Nitschmann 2007-2: 163-167).

From the French societal perspective, the integrity of the individual becomes a public matter and its violation is considered a violation of the "ordre public". As a consequence, the individual consenting to the violation of his bodily integrity makes a disposition – at least partially – of a common object of legal protection (Rechtsgut). Such a disposition would, however, run contrary to the principle that the individual must not have public objects or public interests at his or her disposal. Consequently, the very consent of the individual does not provide grounds for justification in French criminal law as it is in German law.[18] To overcome this dilemma the French legislature has invented the authorization by law ("autorisation de la loi"), which means that the consent of the individual as an expression of the right to self-determination is one element of a legally recognized justification. Consent is, by this legal rule – as opposed to a formal one –, given the power of law.

Given this brief historical examination of the status of the individual, the French concept of consent indicates a strong social relevance or social obligation (Sozialbindung) of the individual that might influence even attitudes toward the handling of elements of the human body. Apropos, concerning the right to self-determination this implies a weakening, a limitation of its impact. The society-oriented logic requires a strong commitment to solidarity or an altruistic engagement in which public interest such as health or security come into play. In the context of biobanks these interests can be reduced to "conservation, circulation, access" (Bellivier and Noiville 2007: 499). Nevertheless, the French conception is "two edged", comprising also the individual itself, with its integrity and dignity as an object of protection.

The penal provisions on the right to self-determination, namely violation of the obligation of consent in the context of removal of bodily elements, reflect the different moral values of the French and German legal systems. While the German draft law on gene-diagnostics envisages a custodial sentence of one year,[19] in France, the penalty is at least five years (art. 1272-4 CSP).

---

[18] For the French conception see Pin 2002, particularly p. 194; also Nitschmann2007b:. 167-170. In Germany the consent is limited by the morals, § 138 StGB.

[19] Entwurf eines Gesetzes über die genetische Untersuchung beim Menschen 2006, §§ 36, 37.

7 The Right to Self-determination and Biobanks?

Regulatory frameworks in France dealing with the rights of individuals must be seen against the background of this general conception of the status of the individual. As in other countries, the "legal landscape" applicable to collection and research of gene-relevant human material is "highly complex, confusing, uncoordinated, and inadequate".[20] The topic is only fractionally dealt with in various pieces of legislation. In France, comprehensive regulations regarding the matter of elements and products of the human body have existed since the 2004 reform of the bio-ethic-regulations[21] and can be found mainly in the civil code, as previously discussed, the public health code and the regulations on data protection. The regulations comprise the actual collections for research purposes, and in particular the handling of genetic information.

Despite the systematic differences between the French and the German civil law, especially in the handling of property, the central question concerning human material is the same: which existing legal rules can be applied?[22] This problem of the protection of the interests of the protagonists, and namely the rights of the donors including the dimension of self-determination, has been solved neither by the 1994 bioethics legislation nor by its reform in 2004.

A systematic search for a legal solution can be conducted in three normative categories: the law of bioethics, of property and of contract. The difficulty of a definitive classification results from the dual nature of the individual as both person and body. Protection for the person is guaranteed in Art. 16 of the civil code, but through consent to the detachment of bodily elements the person becomes a source of substances that may be classified as objects. In the framework of the bioethical law, the person placing bodily substances at the disposal of another has the status of a donor being deprived of the prerogatives of an owner (Bellivier and Noiville 2005: 106). The right to self-determination might have a different shape, depending on where one puts the emphasis.

Contrary to a German thesis, French literature has not departed from the idea that ownership of bodily elements is possible (Bellivier and Noiville 2005: 104-108). This view has been accentuated by the National Ethics Committee which regards the notion of ownership of living elements as an aporia:[23] "The content of a biobank is derived from a voluntary donor. Ownership cannot be transferred to a reasearcher or biobank." Along the same lines, a report of the French Assemblée nationale and the senate points out that genetic characteristics are not comparable to other natural resources. Having the quality of a common good they cannot have

---

[20] For the United Kingdom Gibbons et al 2007: 163, 171.

[21] Loi n° 2004-800 relative à la bioéthique,06/08/2004, Journal Officiel 07/08/2004, 14040.

[22] For Germany see the detailed analysis of Halàsz 2004, for France see Bellivier and Noiville 2006: 101-119.

[23] Comité consultatif national d'éthique 2003.

the status of private or public property.[24] Accordingly, the professionals involved could only function as trustees holding the archive of biomaterial (Bellivier and Noiville 2007: 501).

From a dogmatic point of view, an argument against the property model is that one legal characteristic of an object is that it can have a property value, which, in the case of human material, is prohibited by articles 16-1 and 16-5 of the civil code.[25] Further, the inherent prerogatives of a legal object, the absolute rights of usus, abusus and fructus mentioned in art. 544 of the civil code wouldn't correspond to the handling of bodily elements.[26] Being given as donations and thus following the logic of gifts, any exclusiveness of ownership is thought to be impossible.[27]

Outside of Germany, the right to self-determination doesn't seem to have the same impact as is presumed in German literature in the model of the "Überlagerungsthese". This thesis generally accepts that human elements may constitute property, but the right to property is restricted or even superseded in law by the right to personality (Allgemeines Persönlichkeitsrecht).[28] In light of this, the right to "bio-material self-determination" is discussed as a special category.

In the French discussion on the protection of intellectual property in human biological material, a degradation of the right to self-determination is explicitly deplored.[29] The bioethics legislation of 1994 had previously permitted patents in the field of human materials with respect to innovation beyond the elements or products themselves,[30] and by the revision of these laws in 2004,[31] the French leg-

---

[24] Assemblée nationale/Sénat 2005.

[25] Art. 16-1: Chacun a droit au respect de son corps. Le corps humain est inviolable. Le corps humain, ses éléments et ses produits ne peuvent faire l'objet d'un droit patrimoniale. Les conventions ayant pour effet de conférer une valeur patrimoniale au corps humain, à ses éléments ou à ses produits sont nulles.

[26] Art. 544: La propriété est le droit de jouir et disposer des choses de la manière la plus absolue, pourvu qu'on n'en fasse pas un usage prohibé par les lois ou par les règlements.

[27] The dissection of property is also denied Bellivier and Noiville 2006: 106 f.

[28] For more details see Halàsz 2004: 18 f.: 59-68.

[29] See Bellivier and Noiville 2005: 109-112, Bellivier et al 2006: 288-294.

[30] Loi n° 94-653 relative au respect du corps humain, 29/07/1994, Journal Officiel 30/07/1994, p. 11056.

[31] Loi n° 2004-800 relative à la bioéthique, 06/08 2004, Journal Officiel 07/08/2004, p. 14040; Loi n° 2004-1338 relative à la protection des inventions biotechnologiques, 08/12/2007, Journal Officiel 09/12/2007, p. 20801 Art. L. 611-18 code of intellectual property determines that the "human body in its different states of development, the simple discovery of one of its elements as well as the cognisance of the structure of a human gene as a whole or in parts can't per se be objects of a patent."; for the development on the European level see Convention on the Grant of European Patents, Munich 05/10/1973; Agreement on Trade-Related Aspects of Intellectual Property Rights (TRIPS), Annex 1C Marrakesh Agreement Establishing World Trade Organization, Marrakesh, 15/04/1994, namely Art. 27; Directive 98/44/EC of the European Parliament and of the Council of 6 July 1998 on the legal protection of biotechnological inventions, Official Journal L 213 , 30/07/1998 P. 0013 – 0021.

islature has approved this position. The protection of the dignity of a person from commercial exploitation is guaranteed by legal interdiction: "An invention cannot be patented if its commercial exploitation would be contrary to the dignity of the human person, the "ordre public" or good morals. Such a contradiction is only possible where the exploitation constitutes a violation of a prohibition on legal disposition."[32]

The current situation may be heavily criticized, taking into consideration the intention of medical intervention for industrial purposes and the paradox regarding the profit situation: initial gratitude versus later lucrative exploitation (Bellivier and Noiville 2005: 110).

The legal concept of contract seems to be suited best to the regulation of the different states in order to prevent the removal of human elements from collection to receipt of a patent. Consent as the expression of the right to self-determination appears to be a substantial part of this legal conception. As already indicated, the history of consent in the bioethical context can be described as a constant reinforcement, since its implementation in the course of the legislation on bioethics in 1994.[33] Since 2004, informed consent, referring not only to the act of removal but also to the purpose of utilization, has to precede the removal of any substance, even if, after surgical intervention, the substance removed is used for a research purpose. In the past, materials resulting from an operation were designated "res derelictae" and were free to be used for any purpose,[34] but now they may only be used in regard to blood donations or in experimentation on human beings.

If the intended purpose of utilization of the material changes, the donor must be informed and has the right to oppose the change of use. The status of a donor is compared to that of an author, whose interests must be protected against exploitation by the protection of his intellectual property. Although under strict legal terms the notion of property is denied as far as human elements are concerned, a symbolic parallel could be suggested: the biological materials represent the print of personality.

By strengthening consent as a key contractual element, the legislature has attached importance to the right to self-determination and thus recalibrated the relations between the parties involved in a legal transaction regarding bio-material (Bellivier and Noiville 2007: 501-504, Bellivier and Noiville 2005: 115). Nevertheless, the legislative concept may be criticized for offering insufficient protection to the individual, particularly regarding possible opposition in case of change of intended use (Bellivier and Noiville 2005: 116).

---

[32] See art. 611-17 code of intellectual property: Ne sont pas brevetables les inventions dont l'exploitation commerciale serait contraire à la dignité de la personne humaine, à l'ordre public ou aux bonnes moeurs, cette contrariété ne pouvant résulter du seul fait que cette exploitation est interdite par une disposition législative ou réglementaire.

[33] For an overview see Bellivier and Noiville 2005: 113-116.

[34] Art. 16-3 civil code, art. 1211-2, 1245-2 code of public health; for the development see Bellivier and Noiville 2005: 113 f.

The impact of contracts in connection with the handling of biomaterials is notable in the case of "Greenberg";[35] the decision is supposed to be transferable to the French system for its emphasis on the possible use of the means of contract (Bellivier and Noiville 2005: 118). The judge constitutes that the personal interests of the plaintiffs could, under certain conditions, have been given legal value through the preparation of an agreement, subject to the consent of the donors.

A position similar to property is de facto created by means of a contract, which permits researchers or industries to reserve priority over the use of the material during a certain period, thus generating exclusive possession and an opportunity for research (Bellivier and Noiville 2005: 108). In practice, contracts apply concepts such as "access" and "utilization" or refer to biomaterial as a "patrimoine commune de l'humanitè" (value of common interests for the human race; Bellivier and Noiville 2007, 503), thus avoiding the notion of property.

One has to concede that the traditional role of the contract as a means of exchange is exceeding its original ideology by constituting, in the context of biobanks, an instrument of social regulation (Bellivier and Noiville 2007: 504). Even though contract complements and therefore cannot displace the category of property, contract at least permits an interim solution regarding the partition of interests.

8 Conclusion

Apparently, normative decision-making is still extremely difficult in the field of biomedicine and genetics, as the example of the right to self-determination in France proves once more. There are still many conflicts that cannot yet be solved due to a lack of rational criteria, and it must be tolerated that there is an inherent risk in regard to making such decisions. One may assume that there will never be complete accord between legislation and technical progress.

Overhasty legislation tends to undermine the integrity of law as a reliable normative basis. Established rules may soon loose their impact due to the "normative power of the factual" (die normative Kraft des Faktischen), as proven by the example of the French legislation on cloning.[36] As a result of too frequent revision, the law could appear inflationary and insecure. The use of general terminology that permits interpretation, and is therefore applicable in a variety of cases, is subject to the principle of determinativeness of law, which is especially relevant in

---

[35] Greenberg et al. V. Miami Childrens's Hospital Research Institute, n° 00 C 6779: The plaintiffs ask a "permanent injunction restraining defendants from restricting access to testing for Canavan diseaes through exclusive licensing and/or collection of royalties, and from impeding research on finding a cure or therapies for Canavan disease through enforcement of the patent."

[36] The development describe Bellivier et al 2006: 286-288.

criminal law as nulla poena/nullum crimen sine lege scripta.[37] To avoid negative consequences, the legislature might fall back on the concept of "soft law" in which the object of legislation is still evolving and adapt the final form of legislation as gently as possible respecting the interests in play.

The observation has been made that the spirit of French regulations in the biomedical and genetics fields tends to favour scientific research rather than to limit these activities by emphasising constitutionally protected principles or rights of the individual (Bellivier et al 2006: 275). This observation stands in contrast to the spirit inhered in the bioethics-legislation of 1994. Some of the rules referring to biomedical research and applying to biobanks in the "Code de la santé publique" (public health code) are indeed disputable with regard to the fundamental principles or constitutional rights established in the "civil code".

The coexistence of different codes resulting in an incoherent application of the law is one of the most neuralgic points. Thus specific legislation comprehending the scenarios provoked by the establishment of biobanks is, in the long term, indispensable.

To enhance the guarantee of the right to self-determination, the option of participation must be extended to individuals at the earliest possible stage of the balancing process. The French model of the "États Généraux" as a platform of discussion is an example of a democratic mechanism at a lower level that might be followed.

One must accept that the right to self-determination is a fluid concept that is continually interacting with social habits, practices and morals, and thus also with scientific development. It is apparent that the growing dimensions of genetics and the hopes related to progress in research have an influence on societal expectations that could perhaps lead to a "laxer" handling of the right to self-determination. The restriction of liberties of the individual must be re-examined regularly in light of society's almost magical trust in the overestimated, or at least inestimable, power of a biological and materialistic concept of the human being.[38] A continual balancing of interests is necessary for the establishment and maintenance of the "ordre public", as well as the fixing of limits, namely interdictions regarding activity in the field of genetics.

References

Bellivier F, Noiville C (2005) La circulation du vivant humain: Modèle de la propriété ou du contrat. In: Revet T (ed) Code civil et modèles, pp 101-119. LGDJ, Paris

Bellivier F et al (2006) Les limitations légales de la recherche génétique et de la commercialisation de ses résultats : le droit français. Rev int du droit comp 2: 275-318

Bellivier F, Noiville C (2007) Contrat et vivant. Revue des contrats: 493-509

Bonnin CJ (1812) Principes d'administration publique vol 2. Renaudiere, Paris

---

[37] In Germany this principle is guaranteed in the Grundgesetz by Art. 103 Abs. 2.

[38] In this sense Bellivier F et al 2006, 318.

Chevallier J (1979) Les fondements idéologiques du droit administratif français. In: Variations autour de l'idéologie de l'intérêt général vol 2. Presses Universitaires de France, Paris

Dupont M, Esper C, Paire C (2003) Droit hospitalier. Dalloz, Paris

Ellscheid G (1979) Die Verrechtlichung sozialer Beziehungen als Problem der praktischen Philosophie. Neue Hefte für Philosophie, 17: 37-61

Ellscheid G (1998) Rechtsethik. In: Pieper A, Tunherr U (eds) Angewandte Ethik: Eine Einführung pp 134-155. C.H. Beck, München

Foulquier N (2003) Les droits publics subjectifs des administrés. Émergence d'un concept en droit administratif français du XIX au XXe siècle. Dalloz, Paris

Garraud R (1828) Traité théorique et pratique du droit pénal français vol 5. Sirey, Paris

Garé T (2003) Le droit des personnes. Dalloz, Paris

Gibbons SMC, Kaye J, Smart A et al (2007) Governing Genetic Databases. Challenges Facing Research Regulation and Practice. J of Law and Soc, 34: 163-189

Guigou E (2001) Exposé des motifs, projet de loi relatif à la bioéthique, document n° 3166, 25. Juni 2001. http://www.assemblee-nationale.fr/projets/pl3166.asp. Accessed 13 February 2008

Halàsz C (2004) Das Recht auf bio-materielle Selbstbestimmung. Grenzen und Möglichkeiten der Weiterverwendung von Körpersubstanzen. Springer, Berlin

Harichaux M (2003) Libertés et droits corporels. In: Amson D et al (eds) Le grand oral: Protection des libertés et droits fondamentaux, pp 431-512. Montchrestien, Paris

Heisig C (1997) Persönlichkeitsschutz in Deutschland und Frankreich. Kovac, Hamburg

Jung A (1996) Die Zulässigkeit biomedizinischer Versuche am Menschen. Carl Heymanns Köln et al

Jung H (2005) Wertende (Straf-)Rechtsvergleichung. Betrachtungen über einen elastischen Begriff. Goltdammer's Archiv für Strafrecht, 2-10

Jung H (1988) Biomedizin und Strafrecht. Z für die gesamte Strafrechtswissenschaft: 3-40

Katzenmeier C (2002) Arzthaftung. Mohr Siebeck, Tübingen

Lefebvre G, Terroine A (1953) Recueil de documents relatifs aux séances des Etats généraux. C.n.r.s., Paris

Mason JK, Mc Call Smith RA, Laurie GT (2002) Law and Medical Ethics. Butterworths, London/Edingburgh

Nitschmann K (2007a) Das Arzt-Patient-Verhältnis im „modernen" Gesundheitssystem. Nomos, Baden-Baden

Nitschmann K (2007b) Chirurgie für die Seele? Eine Fallstudie zu Gegenstand und Grenzen der Sittenwidrigkeitsklausel. Z für die gesamte Strafrechtswissenschaft 119: 547-592

Pin X (2002) Le consentement en matière pénale. LDGJ, Paris

Portes L (1954) À la recherche d'une éthique médicale. Presses Universitaires de France, Paris

Schroth U (2004) Medizin-, Bioethik und Recht. In: Kaufmann A, Hassemer W, Neumann U (eds) Einführung in Rechtsphilosophie und Rechtstheorie der Gegenwart, pp 458-483. C.F. Müller, Heidelberg

Schulze-Fielitz H (1988) Theorie und Praxis parlamentarischer Gesetzgebung – besonders des 9. Deutschen Bundestages (1980-1983). Duncker & Humblot, Berlin

Thouvenin D (2001) La construction juridique d'une atteinte légitime au corps humain. Recueil Dalloz, Justices, Hors-série: 113-126

Legislation

France

Loi-Huriet n° 88-1138 vom 20. Dezember 1988 relative à la protection des personnes qui se prêtent à des recherches biomédicales, Journal Officiel vom 22. Dezember 1988, 16032

Loi n° 94-653 vom 29. Juli relative au respect du corps humain, Journal Officiel vom 30. Juli 1994, 11056

Loi n° 94-654 vom 29. Juli relative au don et à l'utilisation des éléments et produits du corps humain, à l'assistance médicale à la procréation et au diagnostic prénatal, Journal Officiel vom 30. Juli 1994, 11060

Loi n° 2002-303 relative aux droits des malades et à la qualité du système de santé, 04/03/2002, Journal Officiel 05/03/2002

Loi n° 2004-800 relative à la bioéthique vom 6. August 2004, Journal Officiel vom 7. August 2004, 14040

Loi n° 2004-1338 relative à la protection des inventions biotechnologiques, 08/12/2007, Journal Officiel 09/12/2007, 20801 Art

Germany

Deutscher Bundestag (2006) Entwurf eines Gesetzes über die genetische Untersuchung beim Menschen 2006. Drucksache 16/3233. Deutscher Bundestag, Berlin

Other documents

Comité consultatif national d'éthique (2003) avis n° 76, A propos de l'obligation d'information génétique familiale en cas de nécessité médicale

Comité consultatif national d'éthique (2003) opinion n° 77, Problèmes éthiques posés par les collections de matériel biologique et les données d'information (2003), www.ccne-ethique.fr/francais/avis/a_077p2.htm#Rapport77_I. Accessed 13 February 2008

Comité consultatif national d'éthique, Nationaler Ethikrat (2003) joint document to opinions n° 77, Problèmes éthiques posés par les collections de matériel biologique et les données d'information (2003) and Stellungnahme Biobanken für die Forschung (2004), www.ccne-ethique.fr/francais/avis/a_077p2.htm#Rapport77_I. Accessed 13 February 2008

Assemblée nationale/Sénat (2004) Office parlementaire d'évaluation des choix scientifiques et technologiques, Rapport sur Les conséquences des modes d'appropriation de vivant sur les plans économique, juridique et éthique (2004), www.assemblee-nationale.fr/12/rap-off/i1487.asp. Accessed 13 February 2008

Assemblée nationale, n° 2046, sénat n° 158, Rapport sur la place des biotechnologies en France et en Europe, 27/01/2005, www.assemblee-nationale.fr/12/rap-off/i2046.asp. Accessed 13 February 2008

Data Protection in Germany

Historical Overview, its Legal Interest and the Brisance of Biobanking

Daniel Schneider

Abstract Data protection law has a comparatively long history in Germany. To understand its meaning for biobanking, it seems important to present the history of data protection from its beginnings in 1970 and its connection with the Population Census Act in 1983 up to the existing legal framework. With reference to the Federal Data Protection Act (FDPA) of 1977 the European influence on the law and the legal status of personal data are discussed. The paper will then look at the current aims of and legal interest in data protection law, including today's question as to how a gradually developed protective standard can fulfill the requirements of a rapidly developing technology and its participants. It will also discuss the distinction between biobanks that have specific research objectives and those projects that, at least initially, do not. Futhermore, a presentation of the legislative status quo in biobanking draws attention to the distinction between aliased data and data rendered anonymous as well as to the legal and practical consequences of such a distinction. The remaining question is whether there is a need for specific statutory legislation in the domain of biobanking or whether the principles of anonymity and coding of dating imposed by general data protection law are sufficient for future major biobanking projects.

1 Why we Have to Protect our Data

Big Brother is Watching You! Anybody who hears this sentence is instantly reminded of George Orwell's book „1984", in which the author describes his nightmare of a perfect autarchy achieved by the excessive use of technical devices. But it is no longer repression by human despots that needs to be feared. Today's threat comes from the "political authority of technology itself, which will certainly find its beneficiaries" (Benda et al 1994, translation by the author). In the mid-18th century, the British philosopher Francis Bacon stated: "Knowledge is power". Knowledge in terms of information about others is an important economic good and thus displays a competitive factor of high economic value (Tinnefeld et al 2005). A modern example of this might be store cards, which are offered widely in order to bind customers to the company and collect information on their personal habits and other data useful for marketing purposes. The danger that these personal data might be handled in a frivolous way is generally underestimated and most of the times accepted unconsciously (Simitis 2007). Thus, we have developed into an information society which is preoccupied with collecting personal data all the time and in all places, making them available through new (mobility-) technology and sharing them with miscellaneous parties even over great distances. This development has an enormous potential, especially for the private sector: while initially fear pointed towards the state as the Big Brother (Simitis 2007), the threat today has shifted to private industry, which now dictates the development of modern data protection law (Bizer 2001).

Over the last couple of years the responsible handling of data in the field of genetic engineering has gained much attention in the area of data protection law: there are only a few debates that are as intense and controversial. With the use of biobanks and the consequent possibility of studying multiple human bodily substances, researchers expect important insights into widespread diseases and the development of effective medication. But among the population there is heavy skepticism concerning the benefits of biobanking. Prospective participants worry that their data may be used for non-medical purposes or for purposes they did not agree to. It is thus not surprising that in a European survey only 42% of German respondents would allow the banking of their genetic information for research. Across Europe, only Austria had lower support (37%) regarding the setup of biobanks (Gaskell et al 2006). What is the cause of this restraint? Is it a lack of trust in the present legal protection of the donor and his donation? So far, there are no specific statutory provisions in Germany that regulate the operation of biobanks for research. In order to protect samples and associated information linked to the donor, it is necessary to investigate whether general data protection in the existing legal framework satisfies both the interests of researchers and the demands of donors.

2 Aims and Legal Interest of Data Protection

A glance at the Federal Data Protection Act (FDPA) suffices to locate the aims of the national German data protection law in section 1 [1] of the FDPA. That provision says: "The purpose of this Act is to protect the individual against the impairment of his right to privacy through the handling of his personal data". The question is how, the "right to privacy" – the object of legal protection –, is to be defined.

The Right to Privacy (RTP) in Germany is a constitutionally guaranteed right. Based on case law it is found in art. 2 [1] and art. 1 [1] of the Constitution.[1] Essentially, the subject of this right is the integrity of personality itself, in the sense that a violation would constitute the person as being parted from his actions. Consequently, the RTP protects an extensive shielded domain of private life, "wherein [the individual] is left to his own devices without observation and is able to communicate with persons of trust regardless of social behavioural expectations" (translation by the author)[2]. The system of constitutional norms reflected in the RTP is the means by which the dignity of the individual and his value as a freely-developing identity are recognised.[3]

But what precisely is the legal right that is constitutionally protected by the relatively loosely formulated term 'privacy' found in the RTP? In addition to self-development and confidentiality (such as utilization of private documents, sexual identity, and intrusion of privacy or the right to anonymity), the RTP applies to one's presentation in public (as for example the right to one's own image) and the protection of the individual's name and reputation (Dreier 2004; Sachs 2007).

For the progress of biomedical research, however, these objects of legal protection are of limited relevance. While the contents and limits of these rights were formerly determined by the so-called "spheres of protection" (i.e. social, private and intimate spheres), the Federal Constitutional Court (FCC) soon recognized that those spheres relate to a certain time, place and individual. Each person understood his private sphere to contain different things; what might be concealed by one might be disclosed by someone else. Therefore, the Court relativized its "theory of spheres" and ordered that each person should have a right to determine both his own public or social role and how to present himself to his particular environment. It is up to each individual, therefore, to evaluate the significance of a breach of privacy in relation to his personality and to decide what his social environment should know about him, especially regarding the influence of modern information and communication technologies. In the field of data protection law, the Court developed the so-called "right of informational self-determination" (RIS) and hence raised the protection of data to a constitutional right. This development, being an aspect of the right to privacy, was a result of a decision of the highest court in Germany, in the context of a particular dispute about the so-called "Decision on Population Census", discussed below. In contrast to the RTP, the RIS represents the creation and preservation of personality rather than the protection of its deformation (Dreier 2004). It protects the individual against unlimited inquiry, storage, use and disclosure of his personal data, thereby guaranteeing the exclusive competence of the individual to make a decision as to the disclosure of his personal data

---

[1] BVerfGE (Federal Constitutional Court Law Reports) 35, p. 202-245, 219 (1973); 72, p. 155-175, 170 (1986); 82, p. 236-271, 269 (1990); 90, p. 263-277, 270 (1994).

[2] BVerfGE (Federal Constitutional Court Law Reports) 90 p. 255-262, 260 (1994).

[3] BVerfGE (Federal Constitutional Court Law Reports) 6, p. 32-45, 41 (1957); 7, S. 198-230, 205 (1958); 34, p. 269-293, 281 (1977)

(Sachs 2007, Schmidt-Bleibtreu et al 2008). As a consequence of the imminent dangers of modern information technology, such as the ready availability of data, its arbitrary transferability and unlimited possibilities of combination, it soon became apparent that evaluation of data on the basis of relevance to the personality would be inadequate. In the early 1980s the FCC decided that irrelevant data no longer exist. Therefore, there is always a risk for the RIS if apparently irrelevant data are combined with one another, for they will always allow for some kind of inference to the person and his personality.[4] No legal order should deny a citizen the right to know what information has been disclosed about him, when and to whom. It is still acknowledged, though, that the degree of data protection should be related to its effect on personal privacy and the indefeasible "core of personality".[5]

Even though the constitutional rights are primarily meant to protect privacy from intrusion by public authorities, they are also of great importance in relation to offences by private parties, which is a consequence of the so called "indirect third-party effect" (Mand 2003, Sachs 2007). As an element of an objective constitutional value system the principles of the RIS reflect the whole legal system and have to be considered at all times (Gounalakis and Mand 1997). For the non-public realm, however, conflicts emerge easily between RIS and the constitutional rights of other parties, such as the right to information (art. 5 [1] of the Constitution) and the right to research (art. 5 [3] of the Constitution). These various rights must also be taken into consideration and the data protection legislation must attempt to balance them with the creation of special rights for scientific research.

3 The History of Data Protection

There has always been sporadic regulation for the protection of privacy (for instance the seal of the confession, doctor-patient confidentiality and the secrecy of mail). Broad and constitutional data protection began in the 1960s in the USA, which had developed the precursors of computer-technology and was the first state to confront the risks of computerization of public life. The "fight for privacy" reached its peak when a committee, chaired by Richard Ruggles (Yale-University), attempted to form a comprehensive National Database to compile information on US citizens in 1965 (Martin and Norman 1972). As a result of impetuous discussions and the fear of a realization of Orwell's vision of "Big Brother" those intentions collapsed. In the end, Senator Edward V. Long, who had insistently lobbied for the protection of individual rights, prevented the National Database and was named "Mr. Privacy" (Bull 1984, Martin and Norman 1972).

---

[4] BVerfGE (Federal Constitutional Court Law Reports) 65, p. 1-71, 42ff (1983).
[5] BVerfGE (Federal Constitutional Court Law Reports) 65, p. 1-71, 45f. (1983).

In Germany, discussions about protection of privacy against automatic data processing began in the late 1960s and the early 1970s when personal rights appeared to be endangered by the establishment of technologies enabling vast and rapid data processing.

3.1 The First Data Protection State Acts

In 1970, the data protection legislation constituted by the German state of Hesse (Data Protection Act of Hesse, DPAH) was not only the first such law in Germany, but also worldwide. This work is seen as a reaction to the plans of the state government to establish a central electronically-controlled database to collect all information on the citizens of Hesse (Moos 2006, Simitis 2007). Initially, the Act comprised only 17 articles defining important rights and instruments that are still in effect today. In 1975, the Land Parliament of Hesse appointed law professor Spiros Simitis to State Commissioner for Data Protection, an office that was then one-of-a-kind-throughout the world (Bölsche 1979). As a supervisory authority under section 7 of DPAH, this office handles claims for rectification of incorrect data (section 4 of DPAH) and is responsible for the prevention of unauthorized access to, and alteration or deletion of, data (section 2 of DPAH). Moreover, under section 3 of DPAH, all people involved in data processing are bound to data secrecy, even after their employment has ended.

By 1981, all the other states in Germany (and by 1992 all the states of the former German Democratic Republic as well) had caught up with their data protection legislation concerning public administration (Moos 2006).

3.2 The Federal Data Protection Act of 1977

Shortly after the first states passed their bills concerning data protection, the federal legislature also initiated consultations. In the meantime, some states had stopped their legislative process, awaiting the federal act in order to achieve federally uniform legislation.

This situation arises out of the fact that Germany is a federal republic. Both the federal and the state governments have legislative competence to implement data protective regulations. The federal government has exclusive legislative power concerning federal administration of data protection as well as state administration regarding the implementation of federal law. Furthermore, the federal government has exclusive competence in regard to data protection concerning intelligence services, issuance of passports, federal civil servants and federal statistics. The federal legislature also regulates criminal prosecution, law concerning aliens, public traffic, the postal system and telecommunication or social law. State governments, by contrast, have autonomous authority concerning the police, municipal affairs, education and culture, as well as the public health system. The situation of state governments observed earlier occurs specifically in those domains in which the powers of federal and state legislature are competing, for example legal procedure, alien right of residence and particularly labor law. Since federal law overrides conflicting state law, there was an attempt to avoid a conflict between the state legislation and federal law from the start.

But the projected FDPA was a long time coming. Between the first draft and its passage at the end of 1976 there was a long period of intensive debate and discussion. When the FDPA was first outlined in 1971 it seemed obvious that the primary focus of the legislation should be data processing conducted by the government itself. It was assumed that legislative protection for data in the private sector was unnecessary, because it was expected that, for cost saving reasons, private business would only store the most important business data (Buchner 2006, Simitis 2007). The legislature soon became aware of the enormous economic benefit that data collection and processing would have for the private sector. It became apparent that huge databases on employees and banking information regarding customer's credit ratings were already in existence. Data protection was therefore regulated under federal legislative competence in both the public sector and the private domains. Consequently, the FDPA of 1977 (FDPA 1977) was an "act of omnibus clauses" governing a broad range of issues in data protection and resulting in its cross-sectional character that could not be categorized in area specific codes (see section 1 [3] of FDPA, Tinnefeld et al 2005, Woertge 1984).

From the start, an axiom of the FDPA was that data processing (all sorts of data and processing methods) always requires enabling legislation or the prior consent of the data subject (i.e. the concerned person) under section 3 of FDPA 1977. In this regard, data protection is based on a fundamental prohibition of data processing in the absence of permission (Ruckriegel 1981). Moreover, it was stated that collection and processing of data by public authorities is generally only permitted if it is absolutely necessary (section 9 [1] of FDPA 1977). A further important principle is the mechanism of self-monitoring by a processing institution, and third party monitoring by the Federal Commissioner for Data Protection (concerning data processing in public administrations) or a supervisory authority in the private sector (see sections 28 ff.; 30; 40 of FDPA 1977). The establishment of a system of sanctions against offences was another protective effect of the FDPA 1977 (sections 41 f. of FDPA 1977).

3.3 The FCC and the Amendment to the FDPA (1990)

A further important mile-stone in the history of data protection law was a noted constitutional adjudication, the so-called "decision on population census", in which the FCC adduced the RIS for the first time under art. 2 [1] and art. 1[1] of the Constitution. The starting point was the Population Census Act of 1983 (PCA), which intended a census of population, occupations, buildings and domiciles, as well as of workplaces. For this purpose it was necessary to collect personal data such as name, address, date of birth, phone number, marital status, religion, source of income and education from everyone registered under the census; additionally, information such as usage of public transport, time of transportation to work-place/training school, hierarchical position in one's place of employment, and even residence in an asylum. In addition to the statistical function, these data were to help correct the existing registers of residents and create a basis for political and scientific decision-making.

There was a strong resistance to this legislation by those who were worried about the government "spying" on the population (Tinnefeld et al 2005). The problem was that even after the anonymization of data (i.e. deletion of name) the information remained traceable to the original donor.[6] Nevertheless, the federal and most state governments regarded the legislation as in accordance with the constitution. After receiving several constitutional complaints, the FCC put a stop to promulgation of the PCA by interim order on April 12th 1983 until a definite decision on its merits could be reached.[7]

After an October hearing, the FCC on December 15th, 1983 decided that several articles of the PCA violated individual constitutional rights in a manner that was both extensive and unjustifiable. The PCA of 1983 was declared unconstitutional on grounds that it infringed the RIS of the appellant, the power of the individual to make autonomous decisions about disclosure and usage of his personal data. However, the RIS is not an "absolute right". The human being is "an individual who develops in the midst of a social community and is dependent on communication [and therefore] has to accept restrictions of his RIS in the case of predominating interests of the society" (translation by the author).[8] Such predominating interests are, for example, the constitutional legal foundation, the proportionality, the organizational and procedural securing of data, the subject's right to information, and the deletion of information on the data subject as soon as possible ("principle of data reduction").[9] The enormous significance of data protection was obvious thereafter: data protection is protection of a constitutional right (Simitis 2006).

This decision showed that the phrase "data protection", as it had been used in the 1970 Act of Hesse and adopted by subsequent acts, was slightly misleading (Bull 1979, Simitis 1971, Simitis 2006). The phrase implies that the data is protected, whereas it is actually the RIS that is the object of protection; the law protects against the consequences of data collection and processing. While people talked about "data saving" at first, the term "data protection" became common and

---

[6] BVerfGE (Federal Constitutional Court Law Reports) 65, p. 1-71, 17 (1983).

[7] BVerfGE (Federal Constitutional Court Law Reports) 64, p. 67-72 (1983).

[8] BVerfGE (Federal Constitutional Court Law Reports) 65, p. 1-71, 44 (1983).

[9] BVerfGE (Federal Constitutional Court Law Reports) 65, p. 1-71, 44 ff. passim (1983).

was adopted by German legislation later on. In fact, the American term "data privacy" seems to be more accurate.

The consequence of the important decision of the FCC was that the FDPA and several pieces of state legislation required amendment in accordance with the Constitution. Finally, these revisions led to the FDPA of 1990, which came into effect on January 1st 1991.

In the 1990 FDPA, amended provisions included the period for acquisition of data, no fault liability and damage claims. The original legislation had instigated norms for the prevention of harmful data processing (which is the correct method, since knowledge of data is generally an irreversible procedure) but had made no provision for compensation in the event of damage. The amended Act also limited the use of data to specific purposes and strengthened the rights of the data subject and the powers of the supervisory authorities. Finally, it brought in special regulations concerning research and the media. The overall purpose of the amendment was to improve the comprehensibility of the Act, which was to be achieved by its reconstruction and additional definitions (Ordemann and Schomerus 1997). Although the emphasis of the 1990 FDPA lay primarily in the private sector, the regulatory distinction between public and non-public sectors remained. In addition to special regulations in the private sector, the RIS was guaranteed in the non-public sector and potentially conflicting constitutionally guaranteed rights of private business, such as the liberty of action (art. 2 [1] of the Constitution), were not to be "unacceptably restrained" (Simitis 2006). As a result of the amendments the provisions of the Act imposed less intensive restrictions on private (commercial) users of personal data. (Buchner 2006, Wind 1990).

3.4 The European Influence on the FDPA (2001)

A further development was the second amendment of the FDPA, as a result of European guidelines. Directive 95/46/EC of the European Parliament and of the Council of October 24th 1995 on the protection of individuals with regard to the processing of personal data and on the free movement of such data (DIR 95/46/EC) aimed at a harmonization of member state legislation concerning data protection. Art. 4 of DIR 95/46/EC states the principle of applicable national law: the law that is applicable to data processing is the law of the member state in which the data processing establishment has its domicile.

DIR 95/46/EC also contains specifications concerning the quality and admissibility of acquired data (art. 6 and 7 of DIR 95/46/EC) and re-emphasizes that the purposes of data processing are to be specific, explicit and legitimate. An "adequate level of protection" is required by the Directive in relation to the transmission of data to a non-EU country (art. 26 of DIR 95/46/EC). "Adequate protection" implies that the controller has adduced adequate safeguards, based on the character of the data and associated risks.

Neither federal nor state legislatures met the deadline for the incorporation of the Directive (October 24th 1998). The states were the first to bring in their amendments, while the federal government postponed theirs, on grounds that the existing regulation was adequate (Christians 2000, Weber 1995). Federal legislators soon amended the 1990 FDPA to European specifications within record time, after the European Commission initiated legal action against Germany for breach of contract in the European Court of Justice. The briefness of the debate and passage of legislation was attributed not only to fear of condemnation, but also to federal government plans for a new and extensive revision of the data protection law. Its intention was more to fulfill the duty of enactment than to revise the FDPA further (Tauss and Özdemir 2000). This was certainly a very strange and unique proceeding in the history of German legislation.

Unsurprisingly, the FDPA of 2001 is a weak Act and considered by many to be inadequate. Its lack of clarity had been admonished by the FCC years ago. The difficulties encountered in formulating and enacting the law are probably a result of the historical development of data protection in Germany. Christians (2000) compared the situation to a "multi-level building in prefabrication, which has slowly grown, but has now become very complex". A new level has to be built, "but with sometimes unsuitable measures" (translation by the author). Nevertheless, the FDPA of 2001 already contained some principles of "modern data protection", such as the principle of data reduction and data economy, protection of data through technical means, a data protection audit and primary regulations concerning mobile storage and processing media (e.g. chip cards) (Gola and Schomerus 2007). Under the principle of data reduction and economy in section 3a, the design and selection of processing systems are to collect, process or use no personal data, or as little personal data as possible. In the research sector, unless individual identity is required for good reason, it should be impossible to trace the data back to the subject, or at least made difficult by anonymization or the creation of an alias. But for the purposes of biobanks the identity of the data subject is usually important information.

Another innovation of the 2001 amendment was its application to the non-commercial private sector; the FDPA was applicable to every data acquisition, processing and usage by private controllers. In addition, "special categories of personal data" having particular sensitivity (see below) entered the regulatory context for the first time.

Although DIR 95/46/EC seemed to contemplate unification of regulations governing the public and private sectors, the 2001 FDPA adhered to the separation. Today there are still no legislative ambitions regarding the proposed fundamental modernization of the FDPA. This may be further evidence of a fading motivation for privacy protection, which – due to the almost ubiquitous "prostitution of data" – has degenerated to a "relic of an antiquated historical epoch" (Simitis 2006, tranlsation by the author).

4 The Legislative Status Quo in Biobanking

Considering this, the outcome of the survey among Europeans regarding their acceptance of biobanking seems even more surprising: as mentioned before, only 42% of the respondents in Germany would allow banking of their genetic information (Gaskell et al 2006; see also above at 1.). Compared to other European countries, the number of supporters in Germany is below the survey average of 58%. The question remains: which standards do biobanks apply regarding data protection and how are data to be protected? Even if human bodily substances are assigned to, and thus become the property of, the researcher, the validity of the RTP still persists (NER 2004). Among those legal interests within the RIS are:

- the right to inspection of the data stored
- the right to correction of false data
- the right to know, or not to know, the results of the analysis (right to information) and
- the revocation of informed consent at any time (as well as an erasure or blocking of data and destruction or delivery of the donation under section 35 of FDPA) (Committee TA 2007).

In the absence of specific statutory regulations (see section 1 [3] of FDPA), the protection of data, and thus the RIS of the data subject, remain a matter of general data protection law. Questions raised include how a biobank is to account for data collected under the existing system, and how human bodily substances are to be subsumed under the legal term "data". Are the existing statutory regulations adequate or is there a need for specific legislation to comply with the demands of historical jurisprudence and European specifications?

Human bodily substances already comprise the largest supply of personal data (Tinnefeld et al 2005). For the purposes of data protection, however, a specific definition of the term 'data' is necessary, as well as clarification as to how these data are to be accounted for in the framework of the existing provisions.

4.1 Personal Data under the FDPA

Since the data protection law in section 1 [1] of the FDPA applies only to "personal data", it is necessary to define the phrase. The statute indicates that personal data is any information concerning personal or factual circumstances of an identified or identifiable individual (the data subject) (section 3 [1] of FDPA). 'Personal data' therefore comprises any detail that permits the identification of the individual (Gola and Schomerus 2007). This actually means that the human bodily substance itself ("traces" such as skid marks, blood spots, finger prints etc) cannot yet be considered "personal data" (Simitis 2006).[10]

Typically, biobanks contain both the donation, meaning biomaterial such as cells, tissues or blood, and information on the donor. The latter is of interest as it may shed light on previous diseases, familial and social situations and environmental characteristics (such as he domicile being close to a power plant) that are revealing in regard to the donor. The uniqueness of biobanks lies in this ambivalence about stored material: on the one hand there is the collection of bodily substances and its analyses, and on the other a large assemblage of information containing personal details about the donor (NER 2004, Wellbrock 2003). Genetic data (as opposed to the donation) are regarded as personal data under the data protection law, because they provide details of the donor's constitution (Schladebach 2003). Accordingly, the explanatory memorandum of the Commission finds that genetic data are, under art. 2 of DIR 95/46/EC, "personal data" for the purposes of the data protection law (European Commission 1992). Therefore, genetic data are essentially subsumed by the data protection law; whether they are also counted as belonging to the special categories of personal data under section 3 [9] of FDPA requires further examination.

4.2 Sensitive Data under section 3 [9] of FDPA

As mentioned earlier, in connection with the implementation of DIR 95/46/EC into national law, a new quality of data entered the FDPA: we now distinguish between personal data and special categories of personal data. The latter are defined under section 3 [9] of FDPA as "information on a person's racial and ethnic origin, political opinions, religious or philosophical convictions, union membership, health or sex life". The collection of this information, which is called "sensitive" data (Simitis 1999; Däubler 2001), requires either permission of the donor with explicit reference to the data of interest (section 4a [3] of FDPA) or special enabling legislation. Accordingly, special standards have been provided for these categories of personal data , contained in FDPA sections 13 [2] no. 1-9; 14 [5] no. 1-2; 28 [6-9]; and 29 [5]. This differentiation of data on the basis of its character supersedes the appreciation of the FCC, which addressed data by its potential for harm in the respective application context[11]. Now, the potential harmfulness of the use of data only results in the data being defined as "sensitive", indicating a need to describe precisely what this term means.

The final list of special categories of data in section 3 [9] of FDPA coincided exactly with that of DIR 95/46/EC. These data were chosen on the premise that

---

[10] Yet, this is disputed: some argue that unless the substantive bodily materials are inherently "data" overall protection is not possible (see for this dispute: Mand 2005).

[11] BVerfGE (Federal Constitutional Court Law Reports) 65, p. 1-71, 41ff. (1983).

they bear a high risk of discrimination and stigmatization that is to be avoided (Dammann and Simitis 1997, Tinnefeld 2001). This categorization has been criticized, in part: according to Simitis (1998; 1999), categorization according to the character of data is futile since the danger for privacy always depends on the context of application.

It seems obvious, though, that genetic data (i.e. personal data of analysis) should be subsumed under "data concerning health" (see art. 8 [1] of DIR 95/46/EC). Interestingly, genetic data is mentioned neither by DIR 95/46/EC nor the FDPA, despite longstanding recognition of the importance of its express inclusion in the Directive in order to achieve its special protection (Simitis 2006). Nonetheless, genetic data may be considered "health-related content of information" under data concerning health (Dammann and Simitis 1997, Ehmann and Helfrich 1999, Schulte in den Bäumen 2007) or as 'data concerning health' under section 3 [9] of FDPA and should thus be treated as sensitive data (Arning et al 2006, Däubler 2001, Schladebach 2003). A determination on this point is not really necessary, since research in the area of biomedicine is always related to the "health-related content of information". Finally, it can be stated that individual test results also qualify as "data concerning health", and therefore a special category under section 3 [9] and subject to all the special requirements of the FDPA (see sections 4a [3]; 13 [2] no. 1-9; 14 [5] no. 1-2; 28 [6]-[9]; 29 [5] of FDPA). The implication of these provisions for researchers is that collection and processing of these (sensitive) data is only permitted when the donor provides consent with explicit reference to the special data according to section 4a [3] of FDPA.

After determining the character of data used in research it remains to be asked which FDPA requirements affect the collection and processing of sensitive data in accordance with protection of the RIS of the donor. An understanding of this requires a clear differentiation between anonymous or anonymized personal data and data without personal information. The opposite of personal data is not anonymous data but data without any personal information (i.e. "absolutely anonymous data", see Roßnagel and Scholz 2000). In contrast, anonymous (or anonymized) data can only be attributed to the subject by the data subject himself. "Anonymous" comes from the Greek word for "nameless" or "nominal unknown" (Mand 2005). Therefore, anonymous data contain at least one piece of personal information that does not identify the individual (Roßnagel and Scholz 2000).

4.2.1 Data Rendered Anonymous

The data protection regulations do not apply to data that is anonymous or that has been rendered anonymous. Thus, rendering data anonymous is an efficient means of achieving data reduction and data economy (FDPA section 3a) and enabling researchers to elude the application of data protection law. Accordingly, section 40 [2] of FDPA states that whenever the research objective permits it, the data shall be rendered anonymous. Anonymization modifies personal data in such a way that the personal information "can no longer or only with a disproportionate amount of time, expense and labour be attributed to an identified or identifiable individual" (section 3 [6] of FDPA). The process of rendering data anonymous is in itself subject to the data protection law, whereas the modified data – in default of personal information – is excluded from the scope of application (Gola and Schumerus 2007; Simitis 2006).

Leaving aside any marginal knowledge that may be gained through the process, it remains to be asked whether it is at all possible to render bodily substances anonymous. Given that absolute anonymization seems difficult (identification is, at least theoretically, always possible by comparison with a reference sample), the so-called "factual anonymization" is sufficient in Germany. Under section 3 [6] of FDPA, this status is achieved by the modification of data so that it can only be attributed to the subject "with a disproportionate amount of time, expense and labour". Whether data can be considered "rendered factually anonymous" depends on the particular case. Does the data processor possess additional knowledge of the individual data, or is he somehow able to acquire it? What technical resources are available to him, and how much work and time is required to attribute the data to the subject? Whether personal data can be attributed to the data subject is, in part, relative to the circumstances and must be evaluated regularly on a case by case basis. If, for example, additional knowledge necessary for identification of the donor is not available to the institution who holds the personal data, but is available to another organization, then, upon transmission of the initially (factually) anonymous data to the second institution, it suddenly becomes subject to data protection law.

There is another difficulty specific to human bodily substances in biobanks: because of the great information content of genetic material, the risk of re-identification seems to be even higher than with other kinds of samples. The non-coding area of the DNA especially, which is used as a "genetic fingerprint", allows a relatively dependable identification of the data subject when compared to a reference sample (Mand 2005, Tinnefeld et al 2005). This problem is even more acute when it comes to rapid progress in the field of biomedicine, such as the technology of DNA-chips. Parallel to the increasing number of data in stock and the more differentiated, faster and cheaper methods of analysis, the risk of re-identification of initially (factually) anonymous data increases (Wellbrock 2003). Finally, the identification of the donor remains a question of probability (Roßnagel and Scholz 2000). The prevailing view, therefore, is that the factual anonymization of data is sufficient to avoid the application of the data protection law, in which event there is no concrete protection of the donor required for the RIS.

4.2.2 Aliased Data

Most of the time, samples for genetic research are only useful if they can be traced back to their donors. Because personal data is an important source of information or because there is a need for further studies (follow-up), to render the data anonymous is often out of the question for biobanking research. Therefore, data protection law (FDPA section 3a; 40 [2]) requires the data to be at least 'aliased', in order to balance the need for data reduction and the potential for identification of the subject. The term "aliasing" is defined in section 3 [6a] of FDPA: "Aliasing means replacing a person's name and other identifying characteristics with a label in order to preclude identification of the data subject or to render it substantially difficult". By creating an alias, the personal data is modified in such a way that tracing the data back to the subject is only possible with the help of a secure code or 'sort key', which cannot be accessed by third parties, or only with disproportionately high effort (Roßnagel and Scholz 2000). Biomedical researchers often engage a trustee to conduct the process, manage the list of donors and handle the sort key. The aliased data is only disclosed to the researchers by the trustee (Committee TA 2007, Wellbrock 2007).

The processing of personal data by the trustee has to be secured. The sort key and aliased data are typically separated on the basis of both organization and location. If a researcher is handling data without using the sort key, the aliased data are to be treated as though they had been rendered anonymous (Arning et al 2006, Mand 2005, Roßnagel and Scholz 2000). The probability of re-identification of aliased data, which is greater than that associated with anonymous data, is dependent on specific opportunities for attribution: on the one side there is the user of the data (the researcher) who does not know the sort key, and on the other side is its keeper (the trustee). While the trustee can still attribute the data to an individual, it remains anonymous to the processor (Roßnagel and Scholz 2000). For the researcher who does not know the sort key, therefore, the data protection regulations no longer apply. Whether the anonymity of the subject is in danger when the data is handled by any other party is determined by the additional knowledge that this party holds or is able to acquire. Therefore, it remains a question of probability.

5 Conclusion and Discussion

De lege lata, research with data that is rendered anonymous or aliased is permitted in the context of specific research projects (Wellbrock 2003). Another question arises, though, as to whether, in the absence of a specific research purpose (e.g. in a general biobanking context), it is sufficient for the protection of a donor's RIS to define genetic data in data protection law as 'sensitive information'.

Given that research with anonymous and aliased data without the sort key falls outside the scope of data protection law, the intensity and controversy surrounding the discussion is somewhat perplexing. Is this much ado about nothing? Maybe the difficulties associated with data protection requirements are caused by practical and theoretical factors that have not yet been considered in this essay. These might include:

- general biobanks with no concrete research objective/purpose store samples for an indefinite time for future use in research projects that have yet to be defined
- it is often not possible to render data anonymous or give it an alias
- networking of biobanks increases the risk of re-identification of subjects
- genetic data might have special characteristics that require special arrangements ("genetic exceptionalism").

The value that a supply of data has for researchers depends on the quality and quantity of samples and especially on additional information that may have been collected on donors (Wellbrock 2007). Therefore, rendering data anonymous may not always lead to the desired research result and aliasing data might not be possible for the specific research design. Another imminent problem is an anticipated increase in the risk of re-identification that will occur when there is a corresponding increase in the volume of samples in those biobanks without specific research objectives. Specific research projects must be strictly differentiated from general biobanks that are without a specified objective (Wellbrock 2003). Major biobanking projects may have difficulties with regard to the principle of research purpose under section 40 of the FDPA, since no specific purpose is initially definable. Thus, it is almost impossible to determine whether data in a specific project allows data to be rendered anonymous or aliased (under section 40 [2] of FDPA). The requirements for informing the donor and asking him for his permission are even more problematic, since obtaining consent for each step of processing is nearly impossible (Arning, Forgó and Krügel 2006). It is yet to be determined whether the interests of the data subjects (especially the RIS) can be protected by law within the context of major biobanking projects, such as those in Iceland, Estonia and England. Rudimentary forms exist in Germany as well (such as "popgen" in Kiel). A new light is being shed on the subject of health data protection as the risk of re-identification increases with broad networking and cooperation among biobanks and the participation of various institutes in different locations (see continuative: Mand 2003).

In addition, biobanking might present a high risk for the RIS in that the samples:

- can easily be collected without the knowledge of the data subject and
- are suitable for a variety of analyses and can therefore answer any question on genetic grounds (Wellbrock 2003).

Closely linked to this is the question of genetic exceptionalism. The particularity of human bodily substances has been a topic of many detailed essays. It is stated that genetic data are highly predictive of many diseases long before the first symptoms appear. Their predictive validity varies, of course, underlining their brisance (Damm and König 2008, Enquete Committee 2002). The data subject might have no idea, and possibly does not even want to know, about such information. But the relevance of genetic predisposition could be important for blood relatives and raises questions about their right to know/right not to know[12] (Wellbrock 2003,

---

[12] The possible consequences of the research results also demonstrate the implications of the RIS, which is indeed no „absolute right" and is therefore not at the holder's exclusive disposal.

Wiese 1991). Others state that genetic data is not appreciably different from common medical data such as blood-sugar level, cholesterol or information from phenotypical analysis (Taupitz 2002). An in depth discussion of genetic exceptionalism is, however, beyond the scope of this article.

In addition, the fear of the "transparent human" is often raised (see for example: Härtel 2007). When the relevant data stem from the individual's privacy and thus from the "core of his personality"[13], as is the case with genetic data, the threat to the RIS seems to be increased and the data involved seems to need even more protection.

The situation in connection with genetic data has frequently been regarded as deficient, and special legislation concerning the processing of such data has been debated at a political level for some time. Damm (2004) provides a good outline on the status quo of the lively discussion. Furthermore, many call for a "secrecy of research" similar to doctor-patient confidentiality in order to protect the data in contexts outside of the research context (Conference of Commissioner for Data Protection 2004, NER 2004, Wellbrock 2004). At present, a draft bill focusing on genetic diagnosis (draft of the Act on Genetic Diagnosis) has been submitted and subjected to public hearing before the Committee on Health of the German Bundestag (German Bundestag 2007). It is proceeding through the legislative process; by now only basic elements have been enacted.

There are several foreign Data Protection Acts that apply expressly to bodily substances in biobanking research and could therefore be used as templates for national legislation in Germany :

- Act on a Health Sector Database of 1998, Iceland (considered the first legal regulation in the context of biobanking)
- Human Genes Research Act of 2001, Estonia, and
- Biobanks (Health Care) Act of 2003, Sweden; the act does not regulate the design and operation of major biobanking projects but rather applies to individual collections that are operated by a professional provider in the Swedish health care system (Morr 2005).

Some preliminary generic materials on the legal framework regarding the design and operation of biobanks have already been developed and discussed (Committee TA 2007, NER 2004, Pommerening et al 2008, Wellbrock 2003, Wellbrock 2007). The draft bill for the Protection of Self-determination in Genetic Diagnosis submitted by the Conference of the Commissioners for Data Protection (2002) is of great importance in this respect, especially sections 25 ff. which concentrate on the permissibility of research for specified purposes. Additionally, the draft bill of the Act on Genetic Diagnosis supplies, in chapter 7, special regulations for scientific research that address the information and consent procedure in detail (see sections 26, 27). It is not yet clear whether the bill will develop into a special data protection act or a special statute of medical law or even whether it will be passed successfully. So far, it has already passed the German Cabinet but still needs ap-

---

[13] BVerfGE (Federal Constitutional Court Law Reports) 65, p. 1-71, 45f. (1983).

proval by the parliament. Until it comes into effect, research on human bodily substances will be evaluated and regulated with the help of the described principles and instruments of the general data protection legislation.

References

Arning M, Forgó N, Krügel T (2006) Datenschutzrechtliche Aspekte der Forschung mit genetischen Daten. DuD 2006: 700-705

Benda E, Maihofer W, Vogel HJ (1994) Handbuch des Verfassungsrechts der Bundesrepublik Deutschland 2nd edn. de Gruyter, Berlin

Bizer J (2001) Ziele und Elemente der Modernisierung des Datenschutzrechts. DUD 2001: 274-277

Bölsche J (1979) Der Weg in den Ueberwachungsstaat. Rowohlt, Hamburg.

Buchner B (2006) Internationale Selbstbestimmung im Privatrecht. Mohr Siebeck, Tuebingen

Bull HP (1979) Datenschutz als Informationsrecht und Gefahrenabwehr. NJW 1979: 1177-1182

Bull HP (1984) Datenschutz oder die Angst vor dem Computer. Piper, Munich

Christians D (2000) Die Novellierung des Bundesdatenschutzgesetzes – Statusbericht. RDV 2000: 4-17

Committee TA (2007) Bericht des Ausschusses für Bildung, Forschung und Technikfolgenabschätzung (18. Ausschuss) gemäß § 56a der Geschäftsordnung, Deutscher Bundestag Drucksache 16/5374, Berlin

Conference of Commissioner for Data Protection (2004) Entschließung der 67. Konferenz der Datenschutzbeauftragten des Bundes und der Länder am 25./26. März 2004 in Saarbrücken zur Einführung eines Forschungsgeheimnisses für medizinische Daten. www.datenschutz.hessen.de. Accessed 11 March 2008

Conference of the Commissioners for Data Protection (2002) Regelungsentwurf zu einem Gesetz zur Sicherung der Selbstbestimmung bei genetischen Untersuchungen des Ad hoc-Arbeitskreises "Genomanalyse" der Konferenz der Datenschutzbeauftragten des Bundes und der Länder. DuD 2002: 150-155

Damm R (2004) Gesetzgebungsprojekt Gentestgesetz – Regelungsprinzipien und Regelungsmaterien. MedR 2004: 1-19

Damm R, König S (2008) Rechtliche Regulierung prädiktiver Gesundheitsinformationen und genetischer „Exzeptionalismus". MedR 2008: 62-70

Dammann U, Simitis S (1997) EG-Datenschutzrichtlinie: Kommentar. Nomos, Baden-Baden

Däubler W (2001) Das neue Bundesdatenschutzgesetz und seine Auswirkungen im Arbeitsrecht. NZA 2001: 874-881

Dreier H (2004) Grundgesetz Kommentar vol 1, 2nd edn. Mohr Siebeck, Tuebingen

Ehmann E, Helfrich M (1999) EG-Datenschutzrichtlinie: Kurzkommentar. Schmidt, Cologne

Enquete Committee (2002) Schlussbericht der Enquete-Kommission Recht und Ethik der modernen Medizin. Media-Print, Paderborn

European Commission (1992) Explanatory Memorandum of DIR 95/46/EC: COM (92) 422 final

Gaskell G, Allansdottir A, Allum N et al (2006) Europeans and Biotechnology in 2005: Patterns and Trends, Eurobarometer 64.3. ec.europa.eu/research/press/2006/pdf/pr1906_eb_64_3_final_report-may2006_en.pdf. Accessed 12 March 2007

German Bundestag (2007) Wortprotokoll Nr. 16/66 of November 7th, 2007. www.bundestag.de/ausschuesse/a14/anhoerungen/066/prot.pdf Accessed 17 March 2008

Gola P, Schomerus R (2007) BDSG – Bundesdatenschutzgesetz: Kommentar. Beck, Munich

Gounalakis G, Mand M (1997) Die neue EG-Datenschutzrichtlinie – Grundlagen einer Umsetzung in nationales Recht, part 1. CR 1997: 431-438

Härtel I (2007) Durch Gendiagnostik zum gläsernen Menschen? Freiheitsrecht in neuer Bewährung. In: Grote R (ed) Die Ordnung der Freiheit : Festschrift für Christian Starck zum siebzigsten Geburtstag. Mohr Siebeck, Tübingen

Mand E (2003) Datenschutz in Medizinnetzen. MedR 2003: 393-400

Mand E (2005) Biobanken für die Forschung und informationelle Selbstbestimmung. MedR 2005: 565-575

Martin J, Normann ARD (1972) Halbgott Computer. Die fantastische Realität der 70er Jahre. BLV Buchverlag, Munich

Moos F (2006) Datenschutzrecht – Schnell erfasst. Springer, Berlin, Heidelberg

Morr U (2005) Zulässigkeit von Biobanken aus verfassungsrechtlicher Sicht. Lang, Frankfurt aM

NER: Nationaler Ethikrat (2004) Biobanken für die Forschung: Stellungnahme. Saladruck, Berlin

Ordemann HJ, Schomerus R (1997) Bundesdatenschutzgesetz mit Erläuterungen, 6th edn. Beck, Munich

Pommerening K et al (2008) Biomaterialbanken – Datenschutz und ethische Aspekte: Generische Konzepte und Realisierung. Medizinisch Wissenschaftliche Verlagsgesellschaft, Berlin (im Druck)

Roßnagel A, Scholz P (2000) Datenschutz durch Anonymität und Pseudonymität – Rechtsfolgen der Verwendung anonymer und pseudonymer Daten. MMR 2000: 721-731

Ruckriegel W (1981) Datenschutzgesetzgebung in der Bundesrepublik Deutschland. Jura 1981: 346-353

Sachs M (2007) Grundgesetz Kommentar, 4th edn. Beck, Munich

Schladebach M (2003) Genetische Daten im Datenschutzrecht – Die Einordnung genetischer Daten in das Bundesdatenschutzgesetz. CR 2003:225-229

Schmidt-Bleibtreu B, Hofmann H, Hopfauf A (2008) Grundgesetz Kommentar, 11th edn. Heymann, Cologne.

Schulte in den Bäumen T (2007) Genetische (Individual-)Informationen aus Sicht des Datenschutzrechts. Bundegesundheitsblatt – Gesundheitsforschung – Gesundheitsschutz 2007: 1- 9

Simitis S (1971) Chancen und Gefahren der elektronischen Datenverarbeitung: Zur Problematik des "Datenschutzes". NJW 1971: 673-682

Simitis S (1990) Sensitive Daten: Zur Geschichte und Wirkung einer Fiktion. In: Schwander I, Brem E, Kramer EA, Druey JN (ed) Festschrift für M.M. Pedrazzini. Stämpfli, Bern

Simitis S (1999) Zur Internationalisierung des Arbeitnehmerdatenschutzes – Die Verhaltensregeln der Internationalen Arbeitsorganisation. In: Hanau P, Heither F, Kühling J (ed) Richterliches Arbeitsrecht, Festschrift für Thomas Dieterich zum 65. Geburtstag. Beck, Munich

Simitis S (2006) Bundesdatenschutzgesetz – Nomos Kommentar 6th edn. Nomos, Baden-Baden

Simitis S (2007) Hat der Datenschutz noch eine Zukunft? RDV 2007: 143-153

Taupitz J (2002) Gentests beim Abschluss von Personenversicherungsverträgen: Problemorientierte Differenzierung oder Methodendiskriminierung? In: RPG 2002: 43-57

Tauss J, Özdemir C (2000) Umfassende Modernisierung des Datenschutzrechtes in zwei Stufen. RDV 2000: 143-146

Tinnefeld MT (2001) Die Novellierung des BDSG im Zeichen des Gemeinschaftsrechts. NJW 2001: 3078-3083

Tinnefeld MT, Ehmann E, Gerling RW (2005) Einführung in das Datenschutzrecht: Datenschutz und Informationsfreiheit in europäischer Sicht, 4th edn. Oldenbourg, Munich, Vienna

Weber M (1995) EG-Datenschutzrichtlinie: Konsequenzen für die deutsche Datenschutzgesetzgebung. CR 1995: 297-303

Wellbrock R (2003) Datenschutzrechtliche Aspekte des Aufbaus von Biobanken für Forschungszwecke. MedR 2003: 77-82

Wellbrock R (2004) Biobanken für die Forschung – Zur Stellungnahme des Nationalen Ethikrates 2004. DuD 2004: 561-565

Wellbrock R (2007) Generische Datenschutzmodelle für Biomaterialbanken – Problemlösungen und offene Fragen. DuD 2007:17-21

Wiese G (1991) Gibt es ein Recht auf Nichtwissen? – Dargestellt am Beispiel der genetischen Veranlagung von Arbeitnehmern. In: Jayme et al. (ed) Festschrift für Hubert Niederländer zum siebzigsten Geburtstag. Winter, Heidelberg

Wind I (1990) Computerwoche of October 10th:8. Munich

Woertge HG (1984) Die Prinzipien des Datenschutzrechts und ihre Realisierung im geltenden Recht. V. Decker, Heidelberg

The Role of P3G in Encouraging Public Trust in Biobanks

Susan Wallace, Bartha Maria Knoppers

Abstract A key element in the success of a biobank is the trust and support of the public. Building this trust is a difficult process. The field of population genomics raises many ethical and societal issues that must be addressed in order for the public to see participation as a trustworthy activity. The Public Population Project in Genomics (P3G) is an international collaboration dedicated to bringing members of the population biobanking community together to share their expertise for the advancement of population genomics research. Through its fundamental principles of promotion of the common good, responsibility, mutual respect, accountability and proportionality, P3G seeks to create tools and resources to assist those involved in biobanks to conduct their research in such a way as to inspire the trust and participation of the public.

1 Introduction

One key element in the potential success of any biobank is the trust and support of the public (Williams and Schroeder 2004). Potential participants need to trust that the research is useful and that they will not be harmed through their participation. The community or population in which the biobank is involved must approve of its plans and its impact. If the biobank is publically funded, community leaders and funders must agree that this effort is worthy of their resources. Finally, society in general must accept that these scientific efforts are beneficial and in the best interest of the public.

Building and maintaining trust is a difficult process. Population based genomics, like other fields of scientific research, has special issues that must be addressed. These include the use of human tissue, confidentiality, and the return of results from DNA studies. As Beskow has noted, in light of a history of research abuses and a continued belief in genetic determinism, "... clarifying the duties of investigators to participants in population-based research in genomics is important." (Beskow 2004) It is therefore vital to assist investigators and researchers to conduct their research to a high standard that will justify the trust of the public.

The Public Population Project in Genomics (P3G) is an international collaboration dedicated to bringing members of the population biobanking community together to share their expertise for the advancement of population genomics research. By creating a repository of resources and tools, by collaborating on research projects and by promoting the sharing of knowledge, it is creating an infrastructure to encourage interoperability in human population genomics as well as a resource for information on best practices in this field. This paper will discuss how P3G, by providing these tools and resources, seeks to help those involved in biobanks to conduct research of high quality that will inspire the trust and participation of the public.

2 Background of P3G

P3G was formally launched, after three years of planning, in May 2007. (P3G 2007a) P3G is not itself a biobank and does not collect data and samples from participants. It works with those involved in biobanking,

... to create, harmonize and share methods, tools and information so as to enhance the design of emerging biobanks and to promote compatibility – between studies – of data (e.g. socio-economic and clinical), samples, and supporting infrastructure (e.g. sample and data-management systems) (Knoppers et al 2008).

In recent years, since the sequencing of the human genome, there has been a move towards "... studying the genetic architecture of complex diseases ..." (Smith et al. 2005). With advances in genetic technologies and the increasing availability of genetic information from individuals, studying gene-gene, gene-disease and gene-environment interactions is now possible. Large cohort studies (prospective studies following a large group of people over years) have been in existence for many years. As well, new large-scale population resources, such as CARTaGENE and UK Biobank, are being created (CARTaGENE 2008, UK Biobank 2007). Access to large and well-characterized sets of data provided by such resources has enabled researchers to identify genetic variants related to health and disease (GAIN Collaborative Research Group 2007, Wellcome Trust Case Consortium 2007). While individual biobanks will undoubtedly produce discoveries regarding health and disease, it is now recognized that by pooling data between population resources, it is possible to investigate not only rare diseases but also common complex diseases, with greater confidence and in greater detail. It is P3G's goal to assist biobanks to share these resources.

P3G comprises members from 25 countries, in three membership categories: Charter, Associate and Individual (P3G 2008e). The founding member biobanks were CARTaGENE (Quebec, Canada), the Estonian Genome Project and GenomEUtwin (involving 8 countries conducting twin studies). Other member biobanks represent Europe, North and South America, Africa, Asia and the Pacific.

International Working Groups (IWGs) of P3G carry out its scientific development work. There are currently 4 IWGs: Social, Environmental and Biochemical Investigations (IWG1); Information Curation and Information Technology (IWG2); Ethics, Governance and Public Engagement (IWG3); and Epidemiology and Biostatistics (IWG4). Research 'Cores', which are independently and externally funded research projects, work with the IWGs and address issues related to biobanking, such as policymaking, health systems research and methods for harmonizing and integrating data. Data and tools created by the Cores are housed in the P3G Observatory (P3G Observatory 2008a), a repository of resources for the use of the community. All resources are publically available.

Underpinning these activities is the P3G Charter of Fundamental Principles aimed at promoting ethical comportment and scientific integrity:

- Promotion of the common good – P3G will optimise the benefits of collaborative research for the benefit of all.
- Responsibility – Protection of the interests of all affected stakeholders including families, groups, populations, researchers and research sponsors is the highest priority. Every effort will be made to respond to the concerns of stakeholders in a timely and appropriate manner.
- Mutual respect – The development and sustainability of P3G is based on responsibility, collaboration, co-operation, trust and mutual respect for others, which includes recognition of cultural diversity and the scientific specificity of the projects involved.
- Accountability – All standards, processes and procedures will be transparent and clear, developed on the basis of consensus, and aim to create best practice in the networking of population genomics resources.
- Proportionality – All research materials (such as data and samples) must be protected to the highest standards of privacy and propriety, while at the same time allowing and promoting the free exchange of ideas, data sharing and openness for the benefit of all. (P3G 2008a)

In adhering to these principles, P3G can work internationally, "… span[ning] critical boundaries across cultures and between legal systems "(Knoppers et al. 2008)

3 Trust Building

When individuals agree to participate in research, they trust that their contribution, such as tissue or information, will be used in an ethical manner. A set of criteria for 'ethical' clinical research suggested by Emanuel and colleagues includes value, scientific validity, informed consent and respect for potential and enrolled subjects (Emanuel et al. 2000). These criteria can also be applied to population research. It is the role of P3G to assist researchers to achieve the highest scientific and ethical standards through the creation of tools for use by the biobank community, to build consensus on best practices in the field, and to transfer that knowledge to the general research community. Specific examples will demonstrate the application of P3G's fundamental principles in these areas. Further, recognition of the necessity and practicability of a platform of scientific communication is reflected in the development of a common lexicon to promote the understanding of key definitions and issues involved in population genomics (P3G 2007b).

3.1 Promotion of the Common Good

P3G is founded on the principle that scientific knowledge is a common good belonging to humanity (Knoppers and Joly 2007). Inspired by the 2002 Statement on Genomic Databases of the Human Genome Organisation (HUGO), which maintains that databases of primary sequences are 'global public goods' (HUGO 2002), the objective of P3G is to ensure that access to genomic information is open to all, and for the benefit of humanity. One of its priorities, therefore, is the creation of tools to foster international interoperability. It encourages, for example, the formulation of broad consent regarding participation in the construction of population biobanking resources; it also affirms the right of members to access, including international access, thereby ensuring that raw data produced by such resources remains in the public domain.

3.2 Optimizing Collaborative Research

P3G, through its IWGs and Cores, works on the many different methodological issues that arise in the creation and running of population biobanks. A large number of samples may be required, for example, as in the population studies noted earlier, to confirm causal links between genes and disease (Smith et al 2005). It can be difficult to share data across studies because while the research question may be the same (whether identical twins who smoke develop a particular cancer at the same rate); the information sought and collected (the quantity of cigarettes smoked) may be collated and recorded in different ways (by the number of cigarettes, the number of packs of cigarettes, etc.). When data is formatted differently, it is not easily shared across platforms. In order to address this problem, P3G is collaborating with other organizations on the DataSHaPER (Data Schema and Harmonization Platform for Epidemiological Research) project. This project is creating a tool that includes

> a comprehensive set of variables that ought ideally to be collected by large epidemiological studies and biobanks whenever the fundamental aim is to undertake general-purpose biomedical research (P3G 2008b).

Such a tool will enable biobanks to harmonize some or all of the data they will collect by allowing them to collect information by reference to the same set of variables, increasing the potential for sharing of information and thus the statistical power of their studies. Hopefully, the tool will also help those designing new biobank infrastructures to 'build in' the ability to share data from the beginning of their study. This should optimize the efficiency of the infrastructure by broadening the scope of research conducted and reducing the number of samples required.

Another project, led by a Core at the UK DNA Banking Network, is the DNA Quantity and Quality Control (Q2C) project (UDBN 2008). Q2C seeks to harmonize DNA measurements among laboratories internationally, thus facilitating the movement of samples between laboratories. As research groups around the world are increasingly sharing DNA, it very important to ensure that measurements are consistent. Such projects should improve the ability of biobanks to collaborate with others.

More generally, the P3G Observatory houses data and tools created by the Cores, as well as information related to population research. The Observatory Study Catalogues contain general information, for example, on 122 new, on-going or completed biobank studies from around the world (P3G Observatory 2008c).[1] It also has a repository of pro forma biobank questionnaires (P3G Observatory 2008b) for use as a reference tool, enabling researchers to compare and contrast the research being conducted by various biobanks.

3.3 Responsibility

It is important to P3G to protect the interests of all stakeholders in biobanks. The IWG3 has accordingly created generic consent materials for population biobanks (Wallace et al 2008), by synthesis of the materials of a group of P3G members. The materials include an information pamphlet and consent form that are to act as templates for identification of issues to be considered by biobanks when preparing their own materials. 'Fill-in' boxes allow the templates to be customized by the designer, according to the structure and needs of the new biobank. The templates are publicly available at the P3G Observatory (P3G 2008c), but their use by P3G members is not mandatory; they are merely tools for the assistance of the biobank community and other interested parties.

Future work of the IWG3 will focus on governance and access issues from the perspectives of stakeholders. The arrangements for governance of a biobank can directly affect participation; systems of ethics oversight and research regulations have been shown to contribute to the level of security that participants feel about the research (Dixon-Woods et al 2007). Access to samples and data is important, both to participants, who wish to be fully informed about the uses to which their contribution is being put, and to researchers, who cannot share without participant consent and mechanisms, such as material transfer agreements, being in place.

---

[1] This number is accurate as of March 2008.

3.4 Mutual Respect

Many organizations around the world are involved in setting standards and creating policies in the field of population biobanking. Working in the area of science and ethics of population genetic databases, to name just a few, are the World Health Organization, the International Society for Biological and Environmental Repositories (ISBER) and the United Nations Educational, Scientific and Cultural Organization (UNESCO) (P3G 2008d). Other projects developing standards for biobank researchers include: Promoting Harmonisation of Epidemiological Biobanks in Europe (PHOEBE), the European Network of Genomic and Genetic Epidemiology (ENGAGE) and the pan-European Biobanking and Biomolecular Resources Research Infrastructure (BBMRI) (BBMRI 2008, ENGAGE 2008, PHOEBE 2008). In light of the number of organizations issuing policies, guidelines and standards, there is naturally a concern that there will be confusion among the scientific, ethics and policy communities as to which policies or standards ought to be followed. As a result, P3G and other groups have considered facilitating discussion among these various bodies in regard to their work, in order to identify opportunities to collaborate, as well as any gaps that may need to be filled or areas of overlap between them. This could help to ensure that existing projects are not duplicated and that potentially beneficial new initiatives are not lost, due to a belief that 'it's already being done.' It is also shows respect for public funds and contributions of samples and data, both of which are limited resources.

A particular question discussed within P3G has been whether adherence to standards developed by the organization will constitute a mandatory requirement for members. Operating on the fundamental principle of mutual respect, which recognizes the cultural diversity and scientific specificity of its member projects, the clear answer is that it has never been the intention of P3G to impose its standards or tools upon its member organizations. All tools are made publicly available, but their implementation by members is not obligatory. However, members have to adhere to the P3G Charter of Fundamental Principles when involved in P3G-related activities.

3.5 Proportionality

While governance is essential for maintaining public trust and participation, controls for the review of research protocols, for data and sample security, and for ongoing monitoring should be proportionate to the sensitivity of the data.

… "Whereas to determine whether a person is identifiable account should be taken of all the means likely reasonably to be used either by the controller or by any other person to identify the said person." This means that a mere hypothetical possibility to single out the individual is not enough to consider the person as "identifiable". If, taking into account "all the means likely reasonably to be used by the controller or any other person", that possibility does not exist or is negligible, the person should not be considered as "identifiable", and the information would not be considered as "personal data" (EC 2007).

Thus, the more open aggregated databases anonymize data. Longitudinal studies, with coded samples that have more phenotypic data attached, require a higher level of security. P3G is constructing a typology of models of data security, access and governance that reflect this proportional approach.

4 Conclusion

The goal of P3G is to help researchers to harmonize aspects of their population genomics research in order to optimise it for the benefit of the public. It is not aimed at diluting that which is special, both culturally and scientifically, about a particular biobank project. Biobanks should differ, according to the needs and questions that will benefit a particular community. But what can be learned from individual communities can also be of use to many others. Where sharing can occur and where harmonisation is possible, P3G will lead efforts to facilitate them. In this way, population biobank research can be used as effectively as possible to best serve the needs of the public. The public trusts that we will do no less.

Acknowledgments The Université de Montréal Policymaking Core is funded by the Canada Research Chair in Law and Medicine, Genome Canada, Génome Québec and the Centre de recherche en droit public, Université de Montréal.

References

Beskow LM (2004) Ethical, legal and social issues in the design and conduct of human genome epidemiology studies. In: Khoury MJ, Little J, Burke W (eds) Human Genome Epidemiology. Oxford University Press, New York
BBMRI (2008) European Biobanks. http://www.biobanks.eu. Accessed 13 March 2008
CARTaGENE (2008) Information Leaflet. www.cartagene.qc.ca/accueil/documents/depliantEn.pdf. Accessed 13 March 2008
Dixon-Woods M, Ashcroft RE, Jackson CJ et al (2007) Beyond "misunderstanding": Written information and decisions about taking part in a genetic epidemiology study. Soc Sci Med 65: 2212-2222
EC (2007) European Commission Article 29 Data Protection Working Party. Opinion 4/2007 on the concept of personal data. http://ec.europa.eu/justice_home/fsj/privacy/docs/wpdocs/2007/wp136_en.pdf. Accessed 17 March 2008.
Emanuel EJ, Wendler D, Grady C (2000) What makes clinical research ethical? JAMA 283: 2701-2711
ENGAGE (2008) European Network of Genomic and Genetic Epidemiology. www.euengage.org. Accessed 17 March 2008.

GAIN Collaborative Research Group (2007) New models of collaboration in genome-wide association studies: The Genetic Association Information Network. Nat Genet 39: 1045-1051

HUGO (2002) Hugo Ethics Committee Statement On Human Genomic Databases. www.hugo-international.org/Statement_on_Human_Genomic_Databases.htm. Accessed 16 March 2008

Knoppers BM and Joly Y (2007) Our social genome? Trends Biotechnol 25: 284-288

Knoppers BM, Fortier I, Legault D et al (2008) The Public Population Project in Genomics (P3G): A proof of concept? Eur J Hum Genet 16: 664-665

P3G (2007a) Canada's new government and Quebec government makes one of the largest investments in international genomics research. www.p3gconsortium.org/news/P3G_pressrelease_English_May192007.pdf. Accessed 5 March 2008

P3G (2007b) P3G Biobank Lexicon. www.p3gobservatory.org/biobankLexicon.do. Accessed 16 March 2008.

P3G (2008a) Charter of Fundamental Principles. www.p3gobservatory.org/download/publications/P3gfundamentalprinciples.pdf. Accessed 13 March 2008

P3G (2008b) DataSHaPER (Data Schema and Harmonization Platform for Epidemiological Research). http://www.p3gobservatory.org/scientificAction.do. Accessed 13 March 2008

P3G (2008c) Ethics, Governance and Public Engagement. www.p3gobservatory.org/ethics.do. Accessed 13 March 2008

P3G (2008d) Summer newsletter. www.p3gconsortium.org/news/P3GNewsLetter4.pdf. Accessed 13 March 2008

P3G (2008e) Membership. 2008 www.p3gconsortium.org/memb.cfm. Accessed 13 March 2008

P3G Observatory (2008a) P3G Observatory. www.p3gobservatory.org/welcome.do. Accessed 13 March 2008

P3G Observatory (2008b) Cross-sectional Questionnaire Catalogue. www.p3gobservatory.org/questionnaireSearch.do?methodToCall=executeGetQuestionnaireCatalog. Accessed 13 March 2008

P3G Observatory (2008c) P3G Observatory Study Catalogue. www.p3gobservatory.org/studySearch.do?methodToCall=executeGetStudyCatalog. Accessed 13 March 2008

PHOEBE (2008) Promoting Harmonisation of Epidemiological Biobanks in Europe. www.phoebe-eu.org/eway/?pid=271. Accessed 13 March 2008.

Smith GD, Ebrahim S, Lewis S et al (2005) Genetic epidemiology and public health: hope, hype and future prospects. Lancet 366: 1484-1498

UK Biobank (2007) UK Biobank Information Leaflet. www.ukbiobank.ac.uk/docs/infoleaflet0607.pdf. Accessed 13 Sep 2007

UDBN (2008) UK DBA Banking Network DNA quantitation project. www.dna-network.ac.uk/DNA+quantitation+project. Accessed 12 March 2008.

Wallace S, Lazor S, Knoppers BM (2008) Consent and population genomics: The creation of generic tools. (submitted)

Wellcome Trust Case Consortium (2007) Genome-wide association study of 14,000 cases of seven common diseases and 3,000 shared controls. Nature 447: 661-678

Williams G, Schroeder D (2004) Human genetic banking: altruism, benefit and consent. New Genet Soc 23: 89-103

Social Issues

Biobanks: Success or Failure?

Towards a Comparative Model

Herbert Gottweis

Abstract In this contribution it is argued that success and failure of biobanks, defined as their capacity to produce value, depends on establishing a system of governance, a mode of ordering that reflects a strategy for pattering a network of interaction that unfolds along a number of different fields, the scientific/technological field, the medical/health field, the industrial-economic field, the legal-ethical and the socio-political field. Presenting a model for the governance, biobanks are described as a network-structure that is not only a research network but a more extensive network that operates through a variety of nodes in different fields from finance to society and bioethical discourse. Bringing order and stability into such a relatively open and not always well-defined network is the key challenge for today´s governance of biobanks. The more ambitious a biobank project is with respect to its envisioned value with respect to research, health and industrial application, the more essential is a balanced management of the multiplicity of the potentially involved factors determining success and failure.

1 Introduction

In recent years biobanks have received much attention as a new key infrastructure and resource for biomedical research and drug development. The task of either transforming existing biospecimen collections into a new genomics research tool or of creating new population based collections is as daunting as it is challenging. The goal of maintaining or creating a biobank typically goes beyond the mere activity of collecting, but aims at creating value through biobanking by creating a specific knowledge value, a value for fostering health, or a specific economic value (Tupasela 2006; Waldby, Mitchell 2006).

Although biobanks have only recently received much public and political attention, they are not something new in the world of medicine and biological research. The systematic collection of human cells and tissues has been done for many years, even dating back to the 19th century, including fixed and processed as well as frozen viable and nonviable material. Millions of tissue samples are being permanently stored, for example, in pathology institutes in many European countries and in North America. Only recently, however, were large patient registries and population surveys initiated that enabled the coupling of biological and genetic data and general patient data (Gottweis, Petersen 2008).

During the last decade, several European countries have started to establish large biobanks with prospective collection of biological material and health data from donors. Most prominently, in 1998, the Icelandic Ministry of Health had announced its plans for the construction of a Health Sector Database on the entire Icelandic population. These plans, initiated by the private company deCODE Genetics, specified how and under what conditions to assemble medical records – and possibly combine them with genetic data and genealogical records for the purposes of tracking the presumed genetic bases of diseases and increasing economizing in the National Health Service (Pálsson et al. 2002). But Iceland's approach towards biobanking was unique in strategy and context. Practices for collecting biological materials and related data differ widely from institution to institution. Enormous potential exists for health research when comparing freshly collected cell and tissue samples to old stored material, provided that the analyses give comparable results (Cambon-Thompson 2004). In several countries, such as in Great Britain, Iceland, Sweden and France, large, well-organized biobanks or tissue repositories already exist that represent large populations. In addition, some countries have large historical tissue collections that go as far back in time as the 1930s. There are good epidemiological registries for most of this period for different diseases, especially in cancer. But also in North America, Australia and Japan, biobanks are seen as a key ressource, and countries such as China or India have begun efforts to establish bio-collections (Gottweis, Petersen 2008).

These strong efforts to maintain, expand, and establish biobanks have good reasons. Today there is a clear realization in the life-science community that the creation of worldwide biobanks networks and cooperation will constitute a crucial step in rebuilding the genomics/postgenomics apparatus of modern biotechnology. The policy vision behind this development is that the exploitation of biobanks and registries is essential during a period when recent improvements of large-scale research in cell and molecular biology will enable new possibilities for health research, knowledge production, and understanding of causes, progression, prognosis and treatment of different diseases (Berg 2001; Cambon-Thomsen et al. 2003). But the justification of biobank projects goes further than just an assertion of their potential value for research. Ultimately, it is argued, biobanks might be an important step towards the improvement and development of preventive, genetic and "personalized" medicine. In fact, in some countries, such as Japan, the biobank projects are seen as "implementation" of the idea of "personalized medicine", understood as the development of new, "tailored" drugs based on the study of diseases and drug side effects, made possible by genetic database research. In other countries, biobanks are defined as machines of innovation policy. In Estonia, for example, the preparation and establishment of the Gene Bank was construed as more than a large research project; it was defined as a strategy for kick-starting a biotechnology industry and thereby of pushing Estonia's post-Soviet economy towards Western standards (Gottweis, Petersen 2008)

But, as many examples from recent biobank development demonstrate, biobanks are neither quickly established nor easily maintained or brought to the status of an integral and useful element of modern biomedical research and development. Biobanks are highly complex and multi-connected networks whose operation depends on a multitude of factors. Thus, the great interest in biobanks, the related, substantial investments, and the expectations connected with them raises the question of what determines the success or failure of a biobank project. This is, of course, a tricky question to answer. But if the purpose of a biobank is to create knowledge based value, health, or economic value, or a combination of all of them – the success of a biobank is determined by the degree to which such value creation can be accomplished. I will argue that the potential value of a biobank project is closely dependent on its governance structure and strategy.

At first glance, biobanks have few areas that are at risk of failure. Collecting disease tissue samples or DNA does not seem to be an activity that should encounter huge problems or difficulties. But this interpretation is not correct. In fact, much can go wrong in setting up a biobank, from the collection of bio-specimens and DNA, to the analysis of these samples and their storage, to the financing of a biobank, its relationship with the public, and its governance structure. Such difficulties can influence substantially the value of a biobank project. This is important to recognize for currently operating biobanks but also for those to be set up in the future, between Bonn, Baltimore, Budapest, Beijing and Bangalore. Thus, creating and operating a biobank necessitates controlling and ordering a multiplicity of interrelated, unstable and shifting factors that potentially determine the success or failure of a biobank project.

2 Towards a Model for Biobank Governance

I argue that success or failure of biobanks or, expressed differently, their capacity to produce value, depends on establishing a system of governance, a mode of ordering (Law 1994) that can be understood as a strategy for patterning a network of interaction that unfolds along a number of different fields: the scientific/technological field, the medical/health field, the industrial-economic field, the legal-ethical and the socio-political field (see figure 1).

For a biobank to function properly, for example, a variety of quality control mechanisms must be established to ensure the scientific validity of the research done with its specimens. This implies, among other things, the proper training of different personnel and ongoing negotiations and exchanges with a variety of administrators of the university where the biobank is located. But this type of interaction alone would not be enough for a biobank to operate successfully. The biobank needs financing, or support by the public, and such needs cannot be taken for granted and require the operators of a biobank set up complex networks of interaction with a number of actors other than scientists and university administrators, such as companies, journalists or patient representatives. It is precisely the establishment of this type of network or, more precisely, this network of ordering that I consider to be crucial for the success of biobanks. Such networks are strategic, established over time, flexible, never stable but shifting, and always in need of stabilization and ordering interventions.

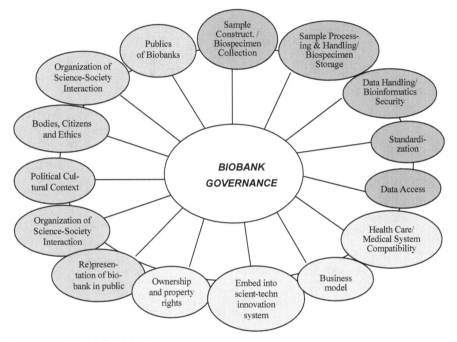

Figure 1: A Model for biobank governance

2.1 Scientific/technical Orderings

Biobanks are research networks (Law 1994). Basically, two large types of biobanks are distinguished in the literature: biobanks based on biological specimens from patients or donors, and population-based research biobanks based on biological samples from (parts of) the general population with or without disease. Population-based biobanks can be cohort studies, whereas subjects are followed over time, case-control or cross-sectional studies. The different biobanks are complementary in the sense that the population-based cohorts depend on endpoints from diagnostic or disease-oriented biobanks, both for precise delineation of phenotypes and for RNA or protein analyses. On the other hand, for etiologic research questions, researchers working with disease-oriented biobanks will need control subjects and biological material that have been collected at an earlier time point as part of the population-oriented cohorts. To date, biobank development worldwide has focused on biobanks based on blood samples (such as in Estonia or Iceland), whereas tissue collections were established in a fragmented manner, resulting in tissue banks of variable size, composition, standards and with different goals (Cambon-Thomsen 2003; Hagen, Carlstedt-Duke 2004; Hirtzlin et al. 2003; Kaiser 2002). A consensus is now emerging, however, that the power of these resources is limited because no single resource contains sufficient samples to cope with biologic/medical diversity. In recognition of the limitations of the current stand-alone biobank model, the establishment of international networks of bio(tissue)banks has been assigned a high strategic priority not only to cover the emerging demands for such resources but also to increase efficacy in medical genomics and to reduce research costs (Bouchie 2004).

Both population-based biobanks and biobanks using biological specimens involve highly complicated, expensive, time-consuming and intricate processes of accessing biological samples, collecting the samples, storing, analyzing and interpreting the samples. While this process is inseparable from society, politics, ethics, law and the economy, it is in itself an enormously difficult operation that requires a high level of scientific competence, technological equipment and know-how, and the diligence to avoid many possibly occurring mistakes.

Typically, bio-repositories collect tissue that was routinely removed for routine medical care such as surgery or diagnostic procedures. This process involves a variety of personnel, such as pathologists, pathology assistants, histotechnologists, tissue technicians and trained repository personnel (Eiseman et al. 2003). Quality assurance is a fundamental component for operating any biospecimen collection (Eiseman et al. 2003, XXI). Developing standardized protocols for a variety of routine acitivities is key in this context. Careful and well-documented processing and extensive annotation of the tissue specimens are crucial for the usefulness of the repository for research (Eiseman et al. 2003). This makes it necessary to train all personnel involved in the collection, processing, annotation, storage and distribution of tissue and to develop standard operating procedures. Also, the relationships between the various types of personnel need to be defined, such as the role of the pathologist to confirm the identity and diagnosis of biospecimens collected by the bio-repository. In this context, developing a bioinformatics system for operating the biobank and protecting data means creating the backbone for the biorepository. Standards for the storage of tissue need to be developed and for monitoring the specimens around the clock. Another key issue is data access and the availability of a bioinformatics system that is searchable and minable via varying levels of web-based access for different stakeholders such as bio-repository personnel, researchers, patients and the public (Eiseman et al. 2003). Each of these steps in handling, describing, and storing biospecimens involves a broad mobilization of knowledge and resources, and each of these steps can go wrong and question the usefulness of the bio-repository under development.

Likewise, population-based biobanks need to carefully build a link between individuals to be part of the planned study and the biobank to be constructed. The identification of potential participants involves, for example, in the case of UK biobank, setting up National Health Service Assessment Centers, creating a standardized protocol to access potential participants, writing questionnaires, sending out invitation letters, training assessment center staff, establishing an assessment center monitoring process, developing a fully integrated IT system, and devising medical check-up and sample collection procedures (Biobank UK 2007). In population-based studies the biobank management must determine an access policy and criteria for prioritization. Just as with bio-specimen collections, population-based biobanks involve extensive preparation and coordination processes to create the kind of research tools envisioned by its creators and funders.

The creation of a biobank as a research network becomes even more complex through increasing cooperation of biobanks internationally that transcend national boundaries. The pan-European Biobanking and Biomolecular Resources Research Infrastructure (BBMRI), a new collaboration of key European biobanks is one example of this trend. Another is the PG3 (The Public Population Project in Genomics) Consortium, whose 'Charter members' include representatives from biobank projects or cohort studies (including samples larger than 10,000) throughout the world. National collections, it is argued, typically suffer from fragmentation of the biobanking-related research community, and with this fragmentation variable access rules and a lack of commonly applied standards for biobanks. This hampers the collation of biological samples and data from different biobanks, which is prerequisite of achieving sufficient statistical power. (Gottweis, Petersen 2008)

Biobank-based post-genome research projects like GenomeEUtwin, an international collaboration between eight twin-registries, typically require strategies for combining extensive amounts of genotype and phenotype data from different data sources located in different countries. This implies not only data harmonization and continuous update of clinical data, but also building infrastructures to provide standardized data exchange and statistical analysis. (Muilu, Peltonen, Litton 2007). In such projects the goal is to facilitate searching, updating, and managing information obtained from various and diverse data sources previously unrelated and located, such as in GenomeEUtwin in various countries in Europe and in Australia.

Thus, biobank projects are always scientific-technological ordering processes in which complicated research networks are created, links between broad varieties of factors are established, and an effort is made to create a certain amount of stability through standardization, normalization, establishment of protocols, data warehouses, databases and identification of strategies for how to link patients/participants with research, analysis and interpretation.

2.2 Networking the Medical/Health Care Field

The goals of many biobank projects are related to a possible translation of biobank projects into clinical practice, drug development and industrial application. At the same time, in many cases, the successful completion of biobank projects depends on the evolution of patterns of cooperation with the health care system, the structure, development level, and practices of a given health care system (Brown, Webster 2004). The lengthy process from procuring tissue samples or blood to analysis, and the application of the results of this research in the health care system, either in the form of new diagnostic possibilities or the development of new drugs, is almost impossible without some friction or conflict. Typically, the research network of a biobank needs to be incorporated or integrated into the medical and the health care system; therefore, the research network biobank must become a broader network stretching to a variety of nodes such as medical schools, hospitals and health care provision. This is not always an easy process, as it involves incorporating and interacting with a broad variety of groups, from geneticists to doctors in different parts of a hospital providing patient records or blood samples.

In Japan, for example, despite backing and financing by the government, Biobank Japan was promoted almost exclusively by private hospital groups rather than by large public university hospitals. When the Biobank Japan project was first announced, members within the Japanese scientific community, including leading geneticists, expressed surprise or even anger with the decision by the Ministry of Education to fund the project. Biobank Japan early on had chosen not to seek contributions from prominent public universities and medical schools. Instead, it enrolled a number of private hospital groups to participate in the project. Engaging medical schools at the University of Tokyo or Osaka University would have meant a complex process of consultation that eventually would have almost certainly led to a fragmentation of project leadership. Keeping sample and information collection simple was a key strategy. Thus, Biobank Japan collects only peripheral blood, rather than tissues, which can be done in a very short period of time. Further, clinical information is partly processed and is entered into a highly standardized format by Biobank Japan's own staff, rather than by the participating hospitals, thus eliminating possible bottlenecks and problems. While doctors need to refer patients to the consultation room, the sampling work is done by the hundreds of medical coordinators at the various hospitals (Triendl, Gottweis 2008).

In Estonia, a different path was chosen to access blood and DNA and to link phenotype (health style and medical record data) and genotype data of 'gene donors' into the gene bank. In the Estonian case (just as in the UK case) the general practitioners (GPs) were to play a key role in linking the biobank project to the health care system and thus the envisioned establishment of a novel system of preventive medicine. The integration of the GPs into the EGP was one of the successes of the project in its early stage. A central motive for their participation in the project seemed to be the prospect of the GPs being part of a prestigious medical-scientific project. Furthermore, the GPs received a strong incentive to participate, as all GPs in the project received IT equipment such as personal computers to be able to process the collected data. In addition, each GP received a financial compensation per donor (Eensaar 2008).

In both the Japanese and the Estonian cases, the crucial importance of embedding biobanks into the larger bio-medical/health network is obvious. This is already a prerequisite for even establishing a biobank and accessing samples and patient information. But such bridging between the research network of the biobank itself and the medical context also constitutes a key resource for translating basic research into application, such as the development of novel drugs, something that is much emphasized as the rationale for many biobank projects. It should not be expected that the medical system or health care providers are necessarily enthusiastic supporters of biobanks. Biobanks potentially constitute significant interventions in the daily life of a hospital or medical school and suggest transformations of health care provision and research practices that are not necessarily welcome by all actors. There are no clear-cut solutions for how to connect and integrate biobanks with the health-case system, but this process is crucial and currently much experimentation is going on in different biobank projects worldwide.

3 Biobanks, Money and Property Rights

Biobank projects tend to be costly projects, ranging from the expenses of maintain an existing bio-repository to staggering investments for setting them up anew, such as in the case of Biobank Japan, which in 2003 received for its initial period of three years a funding of 200 million US-$ (Triendl, Gottweis 2008)

Biobanks potentially create commercial value, or 'biovalue'. Catherine Waldby has interpreted the economic value of biotechnology as biovalue and defined it as "the surplus of *in vitro* vitality produced by the biotechnical reformulation of living processes" (Waldby 2002). Tissues can be leveraged biotechnically so they become more prolific or useful, through processes such as the fractioning of blood, the use of polymerase chain reaction (PCR) for the amplification of genetic sequences, the creation of cell lines, genetic engineering or cell nuclear transfer. The biovaluable engineering is often associated with the requirements for patent so that surplus *in vitro* vitality may eventually be transformed into surplus commercial profits, as well as *in vivo* therapies (Waldby 2006). Not surprisingly, ownership and patenting have become major topics in the discussion on genetic databases. Access, control, and ownership of biobanks and the question of property rights were in particular central issues of those projects in which private industry played an important role (Björkman, Hansson 2006).

Biobank facilities swallow up large amounts of money for the costs of the facility, storage expenses and personnel alone, and these costs run over long periods of time. Such expenses do not support research as such but finance research infrastructures and tools of research; such costs are not easily carried by research funding bodies, universities, or other funding bodies such as research ministries that tend to focus on funding directly basic or applied research. Thus, any biobank needs to have a solid business model that not only provides for its initial set-up but also for its operation over time. It also needs to fit into the national innovation system and its characteristics, such as the availability of venture capital, or the structure of the pharmaceutical industry. Biobank governance, thus, needs to develop a strategy for how to link the research network of a biobank to the worlds of finance, business and state funding.

One key strategy of supporting a biobank is through public financing. This model is at the center of Biobank Japan and UK Biobank, and it has recently also been adopted by the Estonian genome project. Most of the bio-specimen repositories in Europe, such as those located at pathology institutes and hospitals, are also mainly state-funded. In the current discussion, state support for biobanks has been justified by the argument that biobanks can be seen as a tool of international competitiveness. It has been argued, for example, that Europe with its numerous national health care systems seems to have a strong advantage, particularly vis-à-vis the United States, whereas the absence of a national health care system is seen as an obstacle for population-based studies complemented by health data. While governments continue to be major actors worldwide in biobank initiatives, private and non-governmental actors also have come to assume a crucial role. At the same time, in some countries such as Israel and Iceland, narratives of genes as national assets co-exist with privatizing tendencies in biobank development (Gottweis, Petersen 2008).

In Japan in the early 1990s economic arguments were crucial in building political support for personalized medicine and biobank projects and for providing a rationale for the planned huge investment. While similar efforts in the United States or even Iceland such as Celera Diagnostics, Perlegen, or DeCode – were largely funded by private capital raised on the stock market or else financed by the R&D budgets of large multinational pharma companies, no such efforts existed in Japan. Given the crucial role of personalized medicine in the future of healthcare, so it was argued, it was the government who should invest. In other words, public funding for pharmacogenomics and large-scale sample collections in Japan reflected the failure by the Japanese pharmaceutical industry to invest into these emerging technologies. While the limited interaction with industry may well cast a shadow on the long-term future of Biobank Japan (if anything, in the sense that without continuing lobbying, research funding priorities have shifted elsewhere), shielding the project from any corporate influences may well have contributed to its political and public relations success by avoiding some of the complex and difficult debates about ownership, genetic privacy and corporate influences on research thatother biobank initiatives have confronted (Triendl, Gottweis 2008).

Surprisingly little attention has been paid to issues of management, use and ownership of the resources collected. When the the Biobank Japan project was formally announced, some of the most basic questions about ownership were yet to be worked out – and this situation persists to this date. At present, the facilities and resources of Biobank Japan belong to and are owned by the Japanese government but are managed by the University of Tokyo. The situation with respect to intellectual property rights is similar but somewhat less clear, but the absence of clear rules also reflects the fact that companies currently lack access to Biobank Japan. Innovations emerging from the project are patented by the respective organization performing the research. While the project is now approaching the end of its first term, the future of ownerships remains still unclear. Maintaining a sophisticated biobank facility has its cost, and such a move without assurance of continued long-term funding might put the collection in jeopardy. Companies have not been a main target as users for Biobank Japan. Still, uncertain ownership issues have made it only more difficult for companies to actively engage in the project. Companies in Japan have tended to be skeptical about the project and one industry representative interviewed has argued that the 'lack of transparency' at the beginning of the project made it extremely difficult for industry to participate. Unsettled questions about project governance, ownership and long-term strategy have clearly impacted the relationship between the Biobank Japan project and Japanese industry – a fact that appears in striking conflict to the way Biobank Japan has presented itself (Triendl, Gottweis 2008).

The Estonian Genome project, unlike Biobank Japan, developed with a combination of public and private funds. In 2001, the government provided initial funding of 64,000 EUR for creating a public foundation (EGPF), but EGeen, an Estonian company owned by the EGPF and EGeen International Corporation (EGI), which was located in the United States, provided financing for the preparation and establishment of the biobank during 2001-2. In return, the company received an exclusive 25-year commercial license for using anonymous data of the biobank. In 2003 the first conflicts in the consortium began to emerge. EGI said it wanted to concentrate on specific disease groups such as hypertension, and it also cast doubt on the quality of collected data of about 9,000 gene donors and. The conflicts between the EGPF and EGI continued during the next year. Finally in November 2004 the exclusive license and financing contract with EGeen and EGI was terminated, and the Estonian Genome Project (EGP), no longer tethered to a commercial entity, was able to seek public financing. Finally, in 2006 the Estonian state made a decision to support the project in the future (Eensaar 2008).

Although the EGP early on linked the biobank project with public health and established a good connection between societal needs and the operation of the biobank, its initial business model failed soon after the project was launched. The ensuing crisis accentuated the need for a viable business model for any biobank project. Although the EGP was intended as away to launch the Estonian biotechnology industry, this sector was still too nascent for the plan to be effective. In addition, Estonia had little national venture capital prepared to invest in the EGP which forced the EGPF to seek foreign venture capital to invest in a project whose commercial value was unclear (Eensaar 2008).

Iceland's biobank, unlike the previous two examples, was created without government financing, instead being funded by deCODE Genetics, a company founded in 1996 by Icelander Kári Stefánsson and his partner, Jeff Gulcher. The two men initiated the plan for creating a Health Sector Database on the Icelandic population by using the relatively homogeneous Icelandic genome and the expansive local historical records for the purposes of biomedical research.

The project soon created controversy, and the debates within Icelandic society focused on issues of property, ownership and control. Many Icelanders, in particular, found it appalling that a private multinational company would have access to genetic information, medical records and would then explore the genetic bases of common diseases in the Icelandic population and be able to commercialize the results (Palsson 2008). But deCODE Genetics received the license to construct the genetic database in return for a fee paid to the medical service and thus a private model of biobank financing had been created.

In France, yet another model for ownership developed: a patients' organization, Association Française contre les Myopathies (AFM), controlled financing of biobanks. AFM, a private nonprofit-sector patient organization is *the* major actor in the field of biobanking and runs 14 biobanks and collections around the world; in addition, it is active in the areas of genetic research, patient care and legal issues. AFM was created in 1958 with the name of Association Française pour la Myopathie (also called AFM). In 1981 AFM decided to not only foster research but to follow an active research policy by suporting research on *all* genetic disorders and to observe the results. Research on rare diseases, the organization argued, was not sufficiently funded by public research. AFM researchers not only get samples to study but also see the (suffering) patient on a social level. AFM's main goal is to help patients and their families in their daily life, and a major task in this is to break down the invisible barrier between patient and physician or researcher. The two biggest banks of the AFM, both established in the 1990s, are the Banque de Tissus pour la Recherche (BTR) and the Banque d'ADN et de Cellules de Généthon (BG), financed completely by the AFM and governed by the DNA Collection & Banking Department located at Généthon. In 2002, Généthon spent 7% of its 14,6 million Euro budget on its bank and collection department. The mission of the DNA Collection & Banking Department "is to provide the gene therapy research community with human biological sample collections (DNA and tissues)." A charter governs the relationship between the two banks. Both are private, nonprofit disease banks used primarily for research purposes (Mayrhofer 2008).

As these examples show, the presence of a solid business model for biobanks and securing and maintaining the funding, operation, and utilization of a biobank over a long stretch of time are crucial. Although private industry has played a role in recent biobank business models, the cases of Estonia and Iceland demonstrate the difficulties of such forms of support for biobank research networks that do not necessarily yield quick economic returns, yet are based on blood, tissue, and DNA that quickly have raised difficult questions of ownership and benefit sharing.

4 Biobanks, Bodies, Patients, Ethics and Citizens

Biobanks are inseparable from bodies; they study participants in cohort studies and patients donating blood or tissue, and they connect with citizens and society. Establishing and stabilizing the myriad relationships among tissue samples, patient records, blood, DNA- and tissue-donating patients, participants in biobank studies, notions of citizenship, human rights and general understandings of research and medical ethics are indispensable for any biobank project. It is in this respect that a research network links with society and extends into a broader social network structure. Ignoring the complicated social ramifications of any biobank project can quickly lead to serious problems in its operation. A key issue is how a biobank project socially constructs biobank study participants, donors, and citizens. Medical ethics and bioethics are critical discursive resources used in this context.

Unlike in many other medical research projects or studies, the body of biobanks is an inherently decomposed body (Brown and Webster 2004), a body split into systems and collections of blood, proteins, serums, genes, and SNPsAs as opposed to collections of body parts. The elements of the decomposed body in a biobank obtain value through their mutual interconnection. The bodies of biobanks tend to be detached from the persons from whom they originate. The living fluids and living cells of biobanks do not represent other larger bodies but form their own bodies. Biobanks therefore create new 'bodily' phenomenona and new structures for moving bodies and their parts and establishing relationships between them. For example, in the Japan biobank project, the central goals are to assemble blood samples and DNA, to collect clinical information on about 300,000 individuals, based on standard 'informed consent' protocols; to store this information in line with appropriate data safety measures; to determine specific groups of symptoms or reactions to medication using the clinical database and to perform a SNPs analysis covering all genes; and to develop appropriate software tools for the analysis and application of the various datasets created by the project. The goal of the project, therefore, is to create a database that can be used to determine the genetic basis of drug susceptibility and to identify genetic traits related to disease susceptibility, disease progression or responsiveness to certain forms of therapy.

While this goal in itself constitutes a daunting technical-scientific and logistical challenge, it also offers a broad range of ethical and sociopolitical challenges that have given rise to a vast body of ethical-philosophical literature dealing with them. At the core of this challenge is to connect biobanks with individuals in a socially, politically and ethically acceptable manner. This involves mobilizing different narratives that deal with the question of what constitutes a human being, individual and citizen. Bioethicists, philosophers, law scholars, theologians and social scientists engage in this discourse and are engaged in it by the scientists who operate a biobank.

The interrelated topics of informed consent, privacy, autonomy and confidentiality feature highly in this discourse. The idea of protecting the autonomy of the patient/research participant as an individual and citizen armed with political rights and equipped with the capacity to make informed, rational decisions is at the centre of this project. Inherent in this idea is that individuals have the fundamental right to decide about the utilization of their body, body parts, and associated data, and that consent needs to be obtained before any parts or data associated with a particular human body are being used (Cambon-Thomsen, Rial-Sebbag, Knoppers 2007; Shickle 2006).

The idea of informed consent, at first glance, seems to be relatively straightforward. First of all, it is based on the idea of individual autonomy, and ideal of modern culture, the image of the free individual making informed decisions based on personal will and preferences. In the literature, consent procedures are seen as constituting a continuum from highly specific consent to blanket consent. For biobank research, this could mean that an individual either consents to a specific study to be conducted with his/her tissue, DNA and data, that consent could be given to do research on a specific disease, or that consent could be given generally, that is, biomedical research permitting use of the sample for any purpose (Hansson et al. 2006; Porteri; Borry 2008). Furthermore, consent is not something that is given once and then cannot be reconsidered. The Declaration of Helsinki, today´s authoritative statement on biomedical ethics, states that consent of research subjects can be withdrawn at any time without reprisal. Closely related to the field of consent is that of privacy. Over the last decades and in the wake of the Human Genome Project, the concept of genetic privacy and the idea of the need for protection from nonvoluntary disclosure of genetic information has gained prominence (Everett 2004). The principle of the confidentiality of the doctor/patient relationship explains why information from patients needs to be protected from any form of disclosure. The principles of autonomy, informed consent, privacy and confidentiality as basic preconditions for linking people with biobanks have given rise to multi-fold architectures of protocols and standards for how to interact with patients and research participants (Porteri, Borry 2008) and for how to set up sophisticated IT solutions to protect genetic data (Stark, Eder, Zatloukal 2007; Reischl et al. 2006).

These frameworks of creating and dealing with participants and patients in networks of biobanking have begun to be integrated into larger legal structures that create binding rules for the interaction between biobanks and society. Cutter, Wilson and Chadwick have identified two basic models of biobank regulation: first, legislatively created and regulated projects, such as the Icelandic Act on BioBanks establishing the framework for the Icelandic biobank, or the Estonian biobank, projects created specifically by statues and related instruments; second, 'self-created/self-regulated projects, like UK Biobank that are created independently from legislation, and interact with existing laws as the situations arise and that are regulated in a self-binding, but not necessarily legally binding manner' (Cutter, Wilson, Chadwick 2004) In both cases the issues of the establishment and operation of the biobank, the collection, handling, and access to samples, and the relationships with participants, research users and society are regulated through regulations, principles, and ethical guidelines, dealing with issues such as informed consent and confidentiality. Such attempts to govern biobanks develop before, while, or after biobank projects are considered, created and launched, depending on factors such as the pre-existence or the new creation of genetic collections. In both cases, the government or institutional actors close to the state operate as regulators that intend to ensure a sound interaction between biobanks and society. During the last decades, these efforts have not only led to the creation of a legal structure for biobanking but also to the emergence of an international maze of laws, policies, and ethical recommendations that lack harmonization and standardization and constitute an obstacle for research cooperation

As important as these philosophical-theoretical, practical administrative and IT-based efforts are to link society with biobanks, today the reality of developments in large-scale biobanking and genomics has created a highly challenging constellation for biobank governance in which the solutions are less clear than it seems to be in much of the existing literature on the topic. Newly developed technologies such as high-throughput, low-cost sequencing are applied increasingly to human genome and phenome data sets. Comprehensive data sets establish informatics links among genome sequences and extensive phenotype analysis thereby enabling the identification of individuals whose DNA sequence they contain (Lunshof et al. 2008). With the intention of increasing the range and quantity of data, large-scale research platforms are being built that assemble, organize and store data and biospecimens and then distribute them to researchers. Thus, new data flows, genome-wide analysis and novel arrangements between data, patients, and researchers are being established (Lowrance and Collins 2007). Large-scale biobanks are often longitudinal and require extensive exchange of data and specimens implying that a particular sample might be used over the years for varying purposes. All this has given rise to a substantial rethinking of research ethics. As Caplan and Elger argue: "After 50 years of classical health research ethics, regulatory agencies have begun to question fundamental ethical milestones" (Elger, Caplan 2006). In a remarkable paper, Lunshof, Chadwick et al. systematically examine the key concepts of medical research ethics, consent, privacy, confidentiality and then basically abandons these concepts in favor of what they call the concepts of open consent and veracity: "We believe that the building of any comprehensive genotype-phenotype data collection requires that the individuals from whom these data are derived be fully aware that the data can be and likely will be accessed, shared and linked to other sets of information, and that the full purpose and the extent of further usage cannot be foreseen. Individuals should realize that they are potentially identifiable and that their privacy cannot be guaranteed. Full and valid consent by the participants requires veracity on the part of the researchers as a primary moral obligation … Open consent means that volunteers consent to unrestricted redisclosure of data originating from a confidential relationship, namely their health records, and to unrestricted disclosure of information that emerges from any future research on their genotype-phenotype data set, the information content of which cannot be predicted" (Lunshof et al. 2008; cf. Lunshof, Chadwick, Church 2008).

The philosophical justification of this argumentation can be found in the communitarian turn in bioethics, as expressed, for example, in an article in *Nature Review Genetics* written by two of today's leading bioethicists, Bartha Maria Knoppers and Ruth Chadwick, in which they argue that it is time 'to rethink the paramount position of the individual in ethics'. The authors continue in their discussion by approvingly quoting a recent WHO report on genetic databases that states: 'The justification for a database is more likely to be grounded in common values, and less on individual gain. ... It leads to the question whether the individual can remain of paramount importance in this context' (Knoppers, Chadwick 2005, 75). Knoppers and Chadwick then develop their argument by discussing 'new ethical principles' such as reciprocity, mutuality and solidarity as possible strategies to go beyond the more traditional 'individual-centered' approaches in ethics. Thus, the communitarian ideal of the public good seems to overshadow the "classical" ethical orientation towards individual autonomy. It seems that in the face of mounting difficulties in biobank projects to deal with individual-centered approaches in ethics, such as in informed consent, bioethical ideology has already begun to develop a 'new pragmatism' (Knoppers, Chadwick 2005). Hence, linking biobanks with citizens and patients is hardly an easy project that can be built on well-established principles and experiences alone. Scientific and technological advances and the growing importance of cross-national cooperation indicate that well-established medical-ethical procedures that work well, for example, in the field of clinical trials, need to be restructured in the field of biobank research networks. This not only poses a whole range of new theoretical-conceptual issues but also raises the question of the socio-cultural acceptance of these possibly newly arising ways of biobanks connecting with individuals.

5 Linking Biobanks with Society and Publics

Biobanks not only need to be linked to individuals as they participate in cohort studies or donate tissue. A daunting challenge is the question of how to link biobanks with society in general. In the past bio-repositories were leading a quiet life, perhaps in the seclusion of pathology institutes, but with the re-evaluation of existing biobanks and the creation of new ones, multi-faceted medical-ethical issues have arisen along with more general socio-political issues, such as the perception and the acceptance of biobanks in society. While the construction of the participants/patients in biobank projects is a topic of paramount importance, it is important to recognize that most citizens do not belong to this social group. Thus it is a critical element of pattern biobank networks to establish stable links with 'society' and 'publics' as general phenomena. As the recent history tells, biobank projects are not necessarily warmly received in society and even can collapse due to social resistance. In this respect, the political-cultural context of any biobank project plays a very important role.

One problem with 'the public' is that it is not a given entity existing 'out there' so that governments or biobank managers can simply reach out and invite it to participate in some way in the deliberation of a biobank project. Publics of biobanks can spontaneously develop, be shaped through a variety of activists that thematize a particular topic. Such publics emerge from civil society. Participants in these cases have not been 'invited' by government institutions nor have they been selected by formal organizers'; on the contrary, actors are self-selected or 'self-appointed' and as such usually entered the debate from a partisan point of view, promoting their respective cause. Consequently, participation or public involvement that takes place at such unexpectedly politicized sites and is led by civil society rather than by the state tends to feature a rather antagonistic structure, characterized by sometimes adversarial arguments and struggles. Today, participation is often based on the construction of publics by means of a top-down process, such as what could be 'educated publics', that is, formerly 'ignorant' but then 'informed' and 'educated publics', for example constructed via citizen juries. 'Expert publics' are publics composed of actors who are well-informed about a topic to the point that they can claim to represent a topic, such as 'stakeholders' do. Patient groups are an especially important type of stakeholders that either pressure for their participation or involvement in biobank activities or are invited to be represented in various bodies related to biobanks (Gottweis 2008).

In Iceland, an 'unruly', 'spontaneous' public acquired a key role in a biobank project, and, in fact, ended up derailing it. In 1998 the Icelandic Parliament had ratified a bill on a Health Sector Database (HSD) that would assemble in digital form medical records for the entire Icelandic population. s previously mentioned, the company deCODE genetics, which originally outlined plans for the database, was granted an exclusive license for constructing it and using it for 12 years in return for a modest fee, to be returned to the Icelandic community. Polling data showed that the public in general supported deCODE genetics and the database project. In June 1998 a Gallup survey concluded that 58% of Icelanders supported the database, 19% were opposed and 22% were neither for nor against. In April 2000 a Gallup survey concluded that 81% of Icelanders supported the database, 9% were opposed and 10% were neither for nor against. In both 2000 and 2001 the local business magazine *Frjáls verslun* ('Free Trade') declared deCODE genetics the most popular company in the country.

Additionally, the Icelandic stock market responded positively to deCODE Genetics, although the reaction of the market later was mixed and continues to shift (Palsson, Harðardottir 2002).

A strong ethical and political body, Mannvernd – the Association of Icelanders for Ethics in Science and Medicine – was formed in direct response to the database project. Its main spokespersons are physicians, biologists, geneticists and philosophers. Opposition to the project focused on ethical concerns, particularly privacy and the protection of personal medical information. Another important concern was that of informed consent. The HSD was to operate on the basis of the principle of *presumed* rather than informed consent; people could refuse to be included in the collective medical records, but if they did not, information on them would be automatically entered. A fundamental debate took place concerning the ownership of and access to genetic information and medical records. Perhaps the dominant focus was the fact that deCODE, a private multinational company, would have the power to explore the genetic bases in the entire Icelandic population and to then profit from this research. Claims of 'biopiracy', popular in debates surrounding genetic research on indigenous groups and the Human Genome Diversity Project, were present in the Icelandic debate. A major problem in the development of the database project was that a growing number of people opted out of it, refusing to pass on their personal information. By June 2003, roughly 20,000 people had opted out, a significant figure given the size of the population. A further setback was a decision by the Supreme Court in November 2003. The case, *Ms. Ragnhildur Guðmundsdóttir vs. the Icelandic State*, centered on the legality of presumed consent with respect to medical information regarding children, incompetent adults, and the deceased (Supreme Court of Iceland 2003, no. 151). The Court acknowledged the rights of relatives of deceased persons to make decisions about the data involved, thereby adding one more complication to the database project (Palsson 2008). While deCODE genetics had developed an impressive business plan and vision for its biobank project, a gap had widened between Icelandic society and deCODE that from a certain moment in time on was not to be bridged anymore.

The enormous difficulties in developing the Iceland biobank probably prompted the UK biobank's extensive efforts to connect society with the biobank. In 1998, when proposals for a UK genetics population database first emerged, the actors involved in the project realized that ethical considerations would need to be a central concern. The developers of the project determined that they would support a position in which the biobank could be accessed by commercial entities but this access would be subject to adherence to strict ethical protocols. Furthermore, the biobank project would be a public venture funded by UK medical charities and government departments (Corrigan, Petersen 2008).

The funding agencies were aware that the Biobank Project would need to gain the support of not only the half million proposed participants but also the population at large. The Interim Advisory Group (IAG) was created in 2003 and charged with providing a formal ethics and governance mechanism to regulate the UK Biobank project. The IAG was also the advisory body to the UK Biobank's Funders on the best ethical practice to be designed to provide a sound basis for fostering public trust and confidence in the project. The goal of the IAG was as much about minimizing the risk to the project of public rejection and ensuring public trust as it was in minimizing the risks of harm to those participants involved in the research. The Ethics and Governance Council (EGC) is now permanently established to act as an independent guardian of the UK Biobank's Ethics and Governance Framework and to report to 'the public' about the UK Biobank. The project partners have made much of their efforts to 'consult' 'the public' and pertinent stakeholders. These include panels and workshops involving members of 'the general public' from across the UK and specific groups (e.g., people with disabilities or diseases and religious and community groups), meetings with industry and focus groups with primary healthcare workers (Corrigan, Petersen 2008). The first participants were recruited to the UK Biobank in March 2007, and the world's largest planned national repository of human DNA and health-related data for epidemiological research was officially launched.

Apparently, UK biobank had gone through extensive considerations for how to link society with a biobank, and developed a broad range of instruments to implement this process.

6 Conclusions

In this contribution I have argued that success and failure of biobanks, defined as their capacity to produce value, depends on establishing a system of governance, a mode of ordering that reflects a strategy for pattering a network of interaction that unfolds along a number of different fields, the scientific/technological field, the medical/health field, the industrial-economic field, the legal-ethical and the socio-political field. I have described biobanks as a network-structure that is not only a research network but a more extensive network that operates through a variety of nodes in different fields from finance to society and bioethical discourse. Bringing order and stability into such a relatively open and not always well-defined network is the key challenge for today´s governance of biobanks. The more ambitious a biobank project is with respect to its envisioned value with respect to research, health and industrial application, the more essential is a balanced management of the multiplicity of the potentially involved factors determining success and failure.

The different factors of the model I have presented are connected to each other but are not simultaneously equally important and relevant. Bio-repositories of universities might collect disease tissue over long periods of time mainly for internal purposes and not be in need of substantial outside funding nor be under pressure to develop extensive exchange with different social stakeholders. But this situation might change quickly when ambition changes and the mentioned bio-repository begins to develop cooperation with industry or with international research partners. Likewise, public sensibilities towards topics of medical research and development change over time and are different from country to country. Business models that work well in certain periods of time are doomed in others. Nevertheless, I would argue that in the long run all the factors I have discussed in this chapter can play a critical role in the success or failure of a biobank.

The proper scientific-technological standards of a biobank and the establishment of a system controlling its quality might be self-evident in a discussion of the criteria for success, but they nevertheless need to be emphasized. Not all biobanks scientifically and technologically are equally well positioned, and the quality of collections and projects varies within and between countries and regions. Much emphasis has been given to this issue in the literature, for example, in the forms of best practice guidelines and handbooks (Eiseman et al. 2003). Also the issue of developing appropriate medical-ethical and legal frameworks for the operation of biobanks has received much attention in the literature and scientific discussion. But as I tried to show, these issues are shifting continuously in meaning and practice. Rapid progress in genomics research and technology has not only posed pressure on improved collaboration between biobank projects but also raised a range of new issues for ethics and social acceptance. Capitalizing from the new possibilities in research requires investments of unprecedented scope, and the creation of new, large infrastructures in biobanking also raises questions of long-term financial commitment. Solutions for these challenges certainly can be found, but it is important to understand these challenges in their complexity, interactivity, relatedness and dynamism. Only then can the ambition of biobank research be translated into the creation of value for research, health, and industrial application in the medical field.

References

Berg K (2001). DNA sampling and banking in clinical genetics and genetic research. New Genet Soc 20: 59-68

Biobank UK (2007). Protocol for a large-scale prospective Epidemological Resource. Protocol No: UKBB-PROT-09-06 Adswood

Björkman B, Hansson SO (2006). Bodily rights and property rights. J Med Eth 32: 209-214

Bouchie A (2004). Coming soon: a global grid for cancer research. Nat Biotechnol 22, 1071-1073

Brown, N., and A. Webster. 2004. New Medical Technologies and Society Cambridge: Polity Press

Cambon-Thomsen A (2003). Assessing the impact of biobanks. Nat Genet 34: 25-26

Cambon-Thomsen A., Ducournau P, Gourraud PA, Pontille D (2003). Biobanks for genomics and genomics for biobanks. Comp Functl Genom 4, 628-634

Cambon-Thomsen A, Rial-Sebbag E, Knoppers BM (2007). Trends in ethical and legal frameworks for the use of human biobanks. Eur Respir J 30, 373-382

Corrigan O, Petersen A (2008). UK Biobank: bioethics as a technology of governance. In: Gottweis H, Petersen A (eds) Biobanks: Governance in comparative perspective. Routledge, Abingdon: 143-158

Cutter M, Wilson S, Chadwick (2004). Balancing powers: examining models of biobank governance. J Int Biotechnol Law 1: 187-192

Eensaar R (2008). Estonia: ups and downs of a biobank project. In: Gottweis H, Petersen A (eds) Biobanks: Governance in comparative perspective. Routledge, Abingdon: 56-70

Eiseman E, Bloom G, Brower J, Clancy N, Olmsted SS (2003). Human Tissue Repositories. Best Practice for the Genomic and Proteomic Era. Santa Monica: RAND Corporation

Elger BS, Caplan AL (2006). Consent and anonymization in research involving biobanks: Differing terms and norms present serious barriers to an international framework EMBO Reports 7: 661-666

Everett M (2004). Can You Keep a Genetic Secret? The Genetic Privacy Movement. J Genet Counsel 13: 273-291

Gottweis H (2008). Participation and the new governance of life. BioSocieties 3: 265-285

Gottweis H, Petersen A (2008) (eds). Biobanks: Governance in comparative perspective. Routledge, Abingdon

Hagen HE, Carlstedt-Duke J (2004). Building global networks for human diseases: genes and populations. Nat Med 10: 665-667

Hansson MG, Dillner J, Bartram CR, Carlson JA, Helgesson G (2006). Should donors be allowed to give broad consent to future biobank research? Lancet Oncol 7: 266-269

Hirtzlin I, Dubreuil C, Preaubert N, Duchier J, Jansen B, Simon J, Lobato de Faria P, Perez-Lezaun A, Visser B, Williams GD, Cambon-Thomsen A (2003). An empirical survey on biobanking of human genetic material and data in six EU countries. Eur J Hum Genet 11: 475-488

Kaiser J (2002). Population Databases Boom From Iceland to the U.S. Science 298: 1158-1161

Knoppers BM, Chadwick R (2005). Human Genetic Research: Emerging Trends in Ethics. Nat Rev Genet 6: 75-79

Law J (1994). Organizing Modernity. Blackwell, Oxford

Lowrance WW, Collins FS 2007. ETHICS: Identifiability. Genomic Res 317: 600-602

Lunshof J, Chadwick R, Church G (2008). Hippocrates revisited? Old ideals and new realities. Genomic Med 2: 1-3

Lunshof JE, Chadwick R, Vorhaus DB, Church GM (2008). From genetic privacy to open consent. Nat Rev Genet 9: 406-411

Mayrhofer M (2008) Patient organizations as the (un)usual suspects. In: Gottweis H, Petersen A (eds) Biobanks: Governance in comparative perspective. Routledge, Abingdon: 71-87

Muilu J, Peltonen L, Litton JE (2007). The federated database – a basis for biobank-based postgenome studies, integrating phenome and genome data from 600[thinsp]000 twin pairs in Europe. Eur J Hum Genet 15: 718-723

Palsson G (2008). The rise and fall of a biobank: the case of Iceland. In: Gottweis H, Petersen A (eds) Biobanks: Governance in comparative perspective. Routledge, Abingdon: 41-55

Palsson G, Harðardottir KE (2002). For Whom the Cell Tolls: Debates about Biomedicine. Curr Anthropol 43: 271-301

Pálsson G, Harðardóttir KE , Barker JH, Finkler K, Gross M, Helmreich S, Hirsch E, Hornborg A, Metspalu A, Morgan LM, Nelkin D, Proctor RN, Sharp LA, Simpson B, (2002). For Whom the Cell Tolls. Curr Anthropol 43: 271-301

Porteri C, Borry P (2008). A proposal for a model of informed consent for the collection, storage and use of biological materials for research purposes. Pat Educ Couns P 71: 136-142

Reischl J, Schröder M, Luttenberger N, Petrov D, Schümann B, Ternes R, Ternes SS (2006). Pharmacogenetic research and data protection - challenges and solutions. Pharmacogen J 6: 225-233

Shickle D (2006). The consent problem within DNA biobanks. Stud Hist Phil Bio Biomed Sci Part C 37: 503-519

Stark K, Eder J, Zatloukal K (2007). Achieving k-anonymity in DataMarts used for gene expressions exploitation. J Integr Bioinf 4 (Special Issue), http://journal.imbio.de/index.php?paper_id=58

Triendl R, Gottweis H (2008). Governance by stealth: large-scale pharmacogenomics and biobanking in Japan. In: Gottweis H, Petersen A (eds) Biobanks: Governance in comparative perspective. Routledge, Abingdon: 123-139

Tupasela A (2006). Locating tissue collections in tissue economiesderiving value from biomedical research. New Genet Soc 25:33-49

Waldby C (2002). Stem Cells, Tissue Cultures and the Production of Biovalue. Health 6: 305-323

Waldby C, Mitchell R (2006). Tissue economies. Blood, organs, and cell lines in late capitalism. Duke University Press, Durham / London

Sharing Orphan Genes

Governing a European-Biobank-Network for the Rare Disease Community

Georg Lauss

Abstract The majority of current discussions concerning governance challenges in biobanking conceptualize biobanks as technical tools that have to be embedded in a social environment without infringing donors' personal integrity and privacy. This article reconstructs the course of events that induced a European biobank-network operating in the area of rare-disease research: EuroBioBank. Its main focus is on the analysis of social and political complexity involved in the creation of a putatively value-neutral research facility. This case study discusses the governance regime that was adopted to facilitate the intra-European exchange of high quality biological material and its peculiarity in institutionalizing the involvement of organized patient interests. In this model, the participation of organizations of civil society was not regarded as an obstacle, but on the contrary as a valuable resource for effective and democratic governance.

1 Introduction

The last few years have witnessed a significant expansion of collection and processing of human biological samples and of the related information data. Such practices and their institutional contexts are labelled with the term 'Biobank'.

Biobanks are huge repositories of human biological specimens and it is frequently claimed that they have or will have strategic importance for genetic research, clinical care and future treatments. Such endeavours provoke scholarly interest and discussion in the social sciences and in particular among public policy analysts. A great deal of the discussion about governance issues, however, conceptualizes biobanks as bare technical tools or mere resources for scientific research; the suggestion is that they must be properly embedded in a putatively amorphous social environment after their construction.

This analytical framework, which sees the spheres of science and society as two separate domains, occupies the scholarly and public spaces in which guidelines for appropriate policy regimes are formulated. Consequently, most attention is focused on issues such as the adoption of an adequate informed consent procedure and the confidential handling of the data collected. These approaches emphasize the importance of providing the non-scientific public with adequate information and the installation of expert devices that will secure the privacy and confidentiality of the donors (Stark et al 2007).

The aim of this paper is to look at contemporary efforts in biobanks from a different angle. Whereas technology is often regarded as something that exists outside of politics, or even as a way of avoiding the noise and irrationality of a political conflict (Barry 2001), my analysis avoids exclusion of the social and political complexity that is involved in the creation of technological projects. I propose that what is necessary is precisely a better understanding of these complex social and political processes if we want to deal adequately with the challenges of governance associated with the creation of future biobanks and biobank networks.

In a case study I will reconstruct the course of events that led to the construction of the 'European Network of DNA, Cell and Tissue Banks for Rare Diseases' (EuroBioBank or EBB), which is one of the first operating biobank-networks in Europe. The declared aim of this network, which currently consists of 11 relatively small collections of biological material focused mainly on orphan neuromuscular diseases, is to make possible easy access to quality human biological resources for the scientific community. I will demonstrate that this object emerged out of a complex social and political process that is not, and cannot be, restricted to the domain of a value-neutral, techno-scientific expert culture. On the contrary, EuroBioBank evolved out of a set of heterogeneous and contingent political challenges and conflicts. It was induced by the conflation of different interests, practices and types of knowledge that finally created a governance structure that was capable of stabilizing the interactive processes among them.

EuroBioBank evolved out of a political movement of French patient organizations, which had formed a biosocial group to fight for a minority health policy for rare disease patients in France, and subsequently coordinated this transnational European networking process. I will demonstrate that the concept of the 'gene' bound together the interests of a variety of patient organizations and researchers and finally, due to a growing political commitment to promotion of genetic research, the interests of official authorities.

As growing complexity in genomics increased the need for biological samples and associated data, collections concentrating on orphan diseases were among the first to recognize the need for biobank networks. Cases, and therefore biological material, in this area were rare, and sufficient supply could only be provided through the creation of a biobank network that would transcend national borders and operate on the European level. At that moment, the minority health and research policies converged with European integration policy, leading to the involvement of, and financial support by, the European Union.

In the second part of the article I will introduce and discuss the model of private governance adopted in the case of EBB. I will argue that in this model, learning is not understood as a one-way transfer of information or knowledge from experts and bureaucrats to lay people. Rather it creates an institutionalized, reflexive communication process that increases the probability that different epistemic cultures will understand and learn from each other. Civil society organizations are not recognized as an obstacle to policy making, but rather as potential resource for effective and democratic governance. I claim that this interactive and reflexive governance model is particularly suitable for projects like biobank networks, which will only pay off if they are capable of facilitating long-term scientific research projects, thereby operating under circumstances that cannot be fully anticipated under present conditions.[1]

2 Problematizing 'Orphan Genes'

2.1 How to Become a Target for a Political Intervention?

Biobanks are established to facilitate medical and bio-molecular research today, all over the world. This research is increasingly gaining attention, however, because it holds the prospect of developing more reliable diagnostic tools and more effective treatment. In the case of EuroBioBank, the first impetus for its creation came out of the experiences of patients and families of those suffering from a rare disease. For decades, these patients suffered not only from their physical conditions, but from a lack of public and scientific attention to their subject.

At the beginning it was almost impossible to convince state authorities that their problem deserved political attention. By definition, rare diseases affect only a small proportion of the population; as they have a relatively small impact on society, public interest in them is low. Bertrand Barataud, father of a boy suffering from muscular dystrophy, depicts the situation in his book "Au nom de nos enfants":

> Everything was abandonment, resignation, and ignorance. This strategy is found in health care costs, numbers, ethical reasonings, as well as technical and even humanitarian ones. A few bastards evoked such terms as therapeutic excess (Barataud 1992: 14, cf. Rabinow 1999: 38).

This critique was aimed at the political authorities that neglected to recognise rare disease patients as a political problem. And it questions a specific rationality applied by these authorities to determine a fair and just distribution of resources in the domain of public health. 'This strategy' that Barataud found in health care costs, numbers and ethical reasoning was tightly bound to the evolution of the

---

[1] Methodologically I combined qualitative interviewing (Weiss 1994) of a variety of the project's key actors with an anthropological field work approach. I performed this kind of fieldwork at several locations of the network, being it biobanks or patient organizations' headquarters in from April 2007 until November 2007. In addition to this the AFM gratefully provided me with the possibility to analyze the (e-) mail correspondence between several actors that helped me reconstructing the developments since 1998.

modern nation states in Europe in the 19th century. The necessity of knowing a population in order to administer and govern it induced the construction of statistical bureaus (Desrosières 2005:19). The administration's task was to deliver measurements that could be accepted with sufficient congruity to serve as a base for public debates and decision-making (Desrosières 2005:75). This practise of 'objective measurement' found its legitimacy in the overview of the state (Desrosières 2005:57). Barataud and his allies questioned these criteria of justice; criteria that described the treatment that their children needed so desperately as 'therapeutic excess'. "In fact," as Callon puts it, "they expect society to adjust to individual singularities, not the opposite" (Callon 2005: 313). The power relations that governed the health of populations at that time weren't capable of coping with their problems. A new governance regime was needed and some decades later EuroBioBank developed as a substantive part of that regime.

2.2 Making Rare Disease Patients Count

Political conflicts in western societies often rely on the production of numbers that make a problem visible in the public sphere. At that time, no one had compiled statistics on the matter of rare diseases, however, because no one really knew what to count; there wasn't even professional consensus on the description of the phenotypes of most rare diseases. Counting always begins with categorization (Stone 2001:164ff), so how could anybody have started counting? Medical experts were no more interested in the problem than the politicians and bureaucrats. The scarcity of cases itself explained this lack of interest, as it was almost impossible for researchers to access sufficient data to perform scientific research. In France, patients and their families reacted to the situation with the creation of organisations like the "Association Francaise contre la Myopathie" (AFM), founded in 1958. The AFM is a patient network with a specialty in rare muscular diseases, that "decided very early on to include supporting research on its agenda, alongside traditional endeavours such as helping families and fighting for social recognition"(Rabeharisoa and Callon 2006: 143). As a collective, AFM dedicated itself not only to mutual support, but to the accumulation of knowledge on their problem that would facilitate adaptation of the environment of patients and prepare the ground for rare disease research. They created what Rabinow termed a biosocial group (Rabinow 2004: 143) through which both patients and medical experts came to know more about the physical causes of the suffering of rare disease patients. It shaped the identities of patients and blurred the clear demarcation between specialists and patients. Patients were an established part of the highly competent collective (Callon 2005).

2.3 The 'Genetification' of the Problem

By the beginning of the 1980s a plurality of such biosocial groups had been organised, but patient groups of this sort still tended to be heterogeneous, weak and divided (Rabinow 1999: 9). In 1981, the AFM established a high profile scientific council and assigned a primary role to research efforts, which seemed to be the only way towards the development of a cure. Further, as Rabinow pointed out, research increasingly meant genomics. "The 'gene' became the key symbol, the embodiment of fate, the evil locus from which arose death and ruination of innocent life, and, simultaneously, the site of hope" (Rabinow 1999: 39). The gravity of 'the gene', and the hope of cure that it symbolized, managed to order the rare disease community and helped to connect it with the general public interest in genetics. The genetic origin of most orphan diseases became a unifying element for a community representing 6000-8000 different rare diseases. As 80% of these diseases have a genetic origin,[2] projects that promote research in genomics – and therefore potentially treatments and cures – are important connectors that secure the long term stability of the rare disease movement. Patients and scientists started collecting biological materials to make genetic research possible; as we will see, these events resonate quite strongly with the evolution of EuroBioBank.

2.4 Making the Problem 'European'

During the 1990's, patient organizations like the AFM had noticed that in order to draw attention to themselves they needed a critical mass of cases that would be presentable to the public. This critical mass could hardly be achieved at a regional or national level; they needed a voice that could speak for them at the European or even a global level. In 1997, therefore 'The European Organization for Rare Diseases' (EURORDIS) was founded in order to represent the European rare disease patients' organizations. It was conceptualized as a speaking tube with the ability to bundle and amplify the otherwise quiet and diffused voices of its constituency.

Around the same time, the scarcity of high quality biological material was identified as one of the major obstacles to the conduct of research in the area of orphan diseases. The value of small scale biobanks was decreasing, since genetic research could no longer uphold the promise that the 'code of life' could be deciphered by gaining knowledge about the mere qualities of genes (Keller 2000). The hope that stabilized the assemblage, until then, was that they could control health by knowing and controlling the molecular processes that are executed at those special loci on the chromosome. When it turned out that the reality was more complex – that it was impossible to understand most diseases from information inherited by a single

---

[2] EURORDIS, www.eurordis.org/article.php3?id_article=252, 13.02.2008.

gene – regimes to deal with the problem required increased complexity as well. Hope was accordingly concentrated on the creation of biobanks to enable the study of gene-gene and gene-environment interaction. Existing small scale bio-banks could not provide the desired statistical power for that kind of analysis. If available at all, the material was strewn all over the world, where it was stored in the small collections of individual researchers. Often these researchers didn't know about each other, and none of them had enough samples to do effective work with the material they had to hand. The idea that it wasn't enough for a bio-social group to connect voices began, therefore, to gain prominence. In order to find a solution for their genetic problem – to find a cure or a treatment – they had to connect at a European level the already existing collections of body parts such as DNA, cells and tissues.

To summarize, in this section it has been demonstrated that rare disease patient groups sought to be regarded as a political issue. Their success is directly attribut-able to their involvement in research activities and their strategic decision to cou-ple their faith to developments in genetic research. I showed how the creation of EuroBioBank became part of a European integration process. Its establishment re-quired the formation of a European identity, while the European integration proc-ess was at the same time impelled by the successful cooperation of the project.

3 Governing Orphan Genes

The following part of the paper deals with EuroBioBank as a special aspect of general transformations from centralized and hierarchical modes government to more dispersed ways of exercising power. It presents the evolution of a coalition of state authorities and private actors, which is in large part regulated by rules laid down by the involved parties. State authorities constitute one player among others, and no actor has been permitted to dominate the process. This governance model facilitates a process of integrated learning in which scientific experts do not merely inform an amorphous lay public, but empower all involved actors by sup-porting the mutual exchange of information and knowledge. This model is suited to the goal of establishing a transnational network of small scale biobanks, be-cause its flexibility permits the coordinated integration of various actors, including patients and donors, which opens possibilities for mutual understanding, collective learning and the establishment of mutual trust.

3.1 Transformations of Government

Two years after the foundation of EURORDIS in 1997, the EU adopted the 'Program of Community action on Rare Diseases 1999-2003'. The Programme recognized the need for action, and formulated a minority policy capable of tackling the needs of the rare disease activist community. It was clear, however, that an adequate governance regime would have to deviate from the traditional hierarchical and strictly statutory bureaucracy proposed by scholars such as Max Weber (Weber 1976: 825) and Adolf Merkel (Merkel 1927). The European Commission in 1999 stated:

> Rare diseases are considered to have little impact on society as a whole owing to their low prevalence individually; however, they pose serious difficulties for sufferers and their families (Decision No.1295/1999/EC 1999, Article 5).

To address at a political level this matter that was considered by 'society as a whole' to be a minor problem, it was necessary to establish a non-state-centred regime of governance: a policy network. Following Mayntz, I argue that this was not a single event, but part of a larger trend signalling a de facto transformation of political decision-making structures (Mayntz 1993). One of the reasons for this transformation was the growing importance of formal organisations, like the AFM and EURORDIS, which were by then no longer small, divided and non-influential groups of patients, but professional organizations with a considerable amount of financial capacity and know-how. Such organizations are becoming increasingly influential in policy development in almost all societal sectors, leading to an increasing fragmentation of power.

It is important to acknowledge, that centralisation and hierarchical formations of power in modern nation states could develop, because this produced an increased probability to generate collectively binding decisions and stability (Luhmann 2002). It accelerated the speed of communication, and therefore made possible a strategically important gain of time, facilitating a differentiation and improvement of *horizontal* communication. These formally organised and powerful non-state actors, with new communication channels such as the internet enabling them to rapidly coordinate action, are now enhancing the capacity of society to deal with complex social problems in less hierarchical ways. The state is no longer confronted with an amorphous public, but rather with corporate actors that have problem-centred points of view. Interaction between the state and these corporations must be facilitated in order to implement common political objectives (Mayntz 1993). The stability of such networks is secured by their mutually dependent problem-centeredness, meaning that they share interests in objects or areas as well as common goals. In the case under consideration it was 'the gene' and subsequently the social object of a biobank infrastructure that assembled transnational epistemic and practical communities such as patients, scientists, patient activists and public administrators. The EuroBioBank network with its promise of better understanding of gene-gene and gene-environment interaction was a strength drawn upon by the EU in the formulation of its orphan disease policy. The DG Health and consumer Protection of the European Commission (EC) stated that:

> Clearly it is impossible to develop a public health policy, specific to each rare disease. But a global rather than a piecemeal approach could provide some solutions. A global approach to rare diseases means that individual diseases do not fall through the net and real public health policies can be established in the areas of scientific and biomedical research, drug research and development, industry policy, information and training, social benefits, hospitalisation and outpatient treatment[3] (emphasis by the author).

This shared interest in an infrastructure for genomic research made possible 'a global approach' and the creation of a biosocial group of rare disease patients/activists, scientists/geneticists and the European public administration. Together the parties will seek to develop an infrastructure that fosters the optimal use of biological collections through exchange of biological material and associated information.

3.2 EuroBioBank is Taking Shape

In December 2000, the AFM contacted the board of EURORDIS with the following proposal, which was formulated by Francois Salama, a mother of two children who are both affected by rare muscular dystrophies:

> At a time when the Human Genome Sequence is almost completed, coordination and supervision by a European patients organization such as EURORDIS is essential for people suffering from rare diseases to enable them to get involved in their own health and have some kind of contribution and control on what science does for them and with them. You all know the great value of biological collections in rare diseases, and especially DNA or tissue banks, as cases are so few and samples so precious. These samples are essential for research. Thanks to DNA analysis, many diseases due to the defect of a single gene, or several genes, are better understood, many of these genes have been cloned, mutations in patients and parents can be established, a precise, complete and relevant diagnosis can be given, a family specific testing can be set up for genetic counselling, carrier testing and prenatal diagnosis (12/2000 emphasis by the author).

The sequencing of the human genome was almost finished: the first, as it turned out, rather than the last steps towards an understanding of genes, and how they shape our human biological existence. From that point, the importance of biobanks and biological collections for rare disease patient organizations started to increase, for two reasons. Firstly, there was a need for the creation of a resource or tool for translational research on rare diseases in Europe. Secondly, it was important for them that this resource was coordinated and therefore partly controlled by

---

[3] EC: http://ec.europa.eu/health/ph_threats/non_com/rare_diseases_en.htm, 14.02.2008.

organized patient representatives. This strategy framed the patients not as passive objects waiting to be understood by the scientific experts but as active participants in the production of knowledge. One of the first reactions of one of EURORDIS' board members reads as follows:

> I really think that the project is very important for Eurordis and a magnificent opportunity to play a decisive part in the initiative of investigation into rare disorders. … I don't know the position of the other members of the board …, but for the Alliances it is also a new challenge and an opportunity, beyond working in a field, of growing, of *uniting their members* and of collaborating with the investigation. The goal should be to reach, in a shorter time, surer diagnoses and new and effective treatments for the thousands of people that they live in the uncertainty of a sombre prognosis (12/2000 emphasis by the author).

Next to the goal of allegiance with a growing expert culture to develop 'new and effective treatments' and better prognostic instruments, the aim of uniting the members of the Alliances is prominent in that passage. On the one hand, it is important for the organizational rationale to 'speak with one voice' and therefore the stability of the biosocial group; on the other hand it fits neatly with the global approach to rare diseases that was suggested by the EC. Subsequently, the Board of EURORDIS decided to accept the proposed strategy to get involved in the creation of a biobanking network. To prepare a sound financial basis for the network, an application was filed within the 'Quality of Life Programme' of the Fifth European Community Framework Programme covering Research (1998-2002). It was submitted under heading 14: Research infrastructure. The project sought generally to foster user orientation of research, but focused primarily on the scientific aspects. Its main objectives were to: 1) encourage the optimum use of Europe's research infrastructures by fostering transnational cooperation in their rational and cost effective use 2) improve the European wide consistency and complementarity of those infrastructures and their competitiveness at the world level and 3) help to improve quality and user orientation of services offered to the European research community.

Once the EC had approved funding for the project, on the 1st of January 2003, 16 organizations from 8 European countries, among them 12 biobanks, two IT service companies, the AFM and EURORDIS started their work within that framework. The funding was originally provided for 3 years, expiring the 31st of December 2006, but was extended to the 31st of March 2006. At that time Euro-BioBank had already distributed 19,000 out of 170,000 samples listed in its catalogue. Still, coordination and harmonization weren't easy, and the development of conditions, in the form of rules and technical standards, for the coordinated exchange of information, knowledge and biomaterials, required some negotiation and selective exertion of power.

3.3 Identifying Tensions and the Blurring of Traditional Roles in the Policy Process

The common aim of the participants was to share their knowledge and samples. The concept of 'the gene' that had at first connected them was somehow transformed into or enriched by the uniting goal to create EuroBioBank. They all agreed that it was important to understand the complex interplay of genes and environment in the evolution of orphan diseases, and shared the idea that enhanced scientific understanding of that matter could lead to better means for treatment and prognosis. The difficulty was reaching agreement as to how the exchange should take place, without violating the various cultural practices that converged in EuroBioBank. The first challenge was to bridge the gap between scientists' and patients' interest. One representative of the patient organizations told me:

> At the beginning, the creation of this network was a bit of a crazy adventure! Because, even if everybody thinks that it will be very important to exchange biological resources, especially in rare disorders domain, nobody wants to give his 'treasure'!! (05/2007).

The scientists had worked for years to establish collections that were useful for their work. They had invested significant proportions of their time, as well as their financial budgets, to create them. It was no wonder that they were hesitant to give away what they had in their freezers and closets without assurance that they would get something in return. It took time to convince all parties involved that they could come to an agreement:

> They [the scientists] were very reluctant to share all their knowledge ... in particular the samples, of course. Then they understood that we didn't want to steal their ownership. The catalogue wasn't meant to steal whatever they had in the closet. It was just a real tool to ... useful for themselves (05/2007).

The scientists didn't want to give away entire control over their samples, so they agreed to develop a catalogue that would list what they were willing to share. Everything that was listed in the catalogue could be requested from the partners, by any institution actively involved in rare disease research. The terms of exchange, however, were determined by agreement between the involved institutions in regard to each individual transaction. Agreement to this solution was not however very discerning on the part of the scientists, who compete with one another for third party funds and publications. The scientific coordinator of EuroBioBank referred, for example, to "soft pressure from the patient organizations" who knew that scientists are substantially dependent upon them for further donations of fresh biological material and financial support.

The second problem was to deal with cultural tensions between scientific expertise and the bureaucracy. A representative of EURORDIS explained this challenge to me in an interview:

> They [the EC] are used to their system and they don't know how to explain their system to naïve people. ... You are lost in forms, documents, reports and you have the impression you are writing the same thing several times from different angles (05/2007).

The scientists complained about this situation a lot in the beginning. They didn't like to be bothered with that 'unscientific, unpleasant stuff'. It was simply not the way they were used to working, and applied a system of control that was different to the one in their professional domain. It was the patient organisation EURORDIS in its role as project coordinator that established a balance between the different demands. Apparently the EC tried to listen to the criticism; as the project coordinator told me:

> [...] but I don't know how they can simplify it in reality. It is still a big institution with a lot of money to manage. ... They have to put in place a system that can be easily controlled. I understand that from their point of view, but still for scientists who are not used to that mentality it is really tough (05/2007).

The desire of EURORDIS to control, at least in part, the rare-disease policy-making process had led to its involvement as a coordinator of a scientific project. As a result of this, its central task of representing the rare disease community was complemented with other missions. To ensure that project members did justice to the language and administrative system of the EC, it provided translators and facilitators to the Commission's bureaucrats and scientific experts. As coordinator, therefore, the staff of EURORDIS also helped researchers with administrative tasks until they were capable of handling them completely on their own. Through these interactions, patients, scientists and bureaucrats learned from each other and developed a mutual trust and understanding that was to be important in handling the social and technical complexity involved in the activities associated with the creation and direction of a biobank.

In this process, traditional role models became blurred. The following statement of a scientist, for example, explains the importance of databases in the rare disease policy-making process:

> The databases are important in order to tell the government that even if, for example diseases like Duchenne, are rare diseases, there are a significant number of affected patients across Europe, and that of that number a considerable proportion is over 15, and needs a wheel chair and artificial respiration. Or you can put numbers at the politicians' disposal that tell them that at the moment, patients on average grow to 25 years old, and because they are in need of care for 24 hours a day their illness leads to considerable costs for the public health system. Here the whole thing enters a political layer. If you don't have those numbers politicians or public health economists can sit back and tell you that your problem is only something very rare and that it is much more important to do something for cardiovascular diseases or cancer, where they have numbers. I agree that those are important things, too. But we need a certain basis to make progress in our area. That is important for the patient organisations as well (06/2007).

What role does this person play in the policy process? Is it still that of a non- involved expert, who is interested in the production of objective knowledge? I think it is apparent that this is not the case. As I argued in section 2.1 it is first and foremost the task of specialised public administrations to provide the political layer with facts and figures about the population to be governed. In the case of EuroBioBank this task is now partly carried out by a transnational epistemic community, whose members are learning together and from one another. The politics of databases and statistics has left the realm of the state and is now performed by a transnational network of organizations which are simultaneously fulfilling the roles of scientist, activist, administrator and politician. This has made possible a faster and better-coordinated gain of information, thereby reducing the degree of mutual irritation. In 2006, by the end of the EC's funding period, the partners had established a seminal working relationship, which was described by the administrative coordinator of the network:

> In the end, almost all of them understood how a project works and the mentality of the European bureaucrats. After the third year I think they had learned how to answer to the requirements. ... We have established a good relationship that wasn't there at the very beginning. ... Now they [the project partners] have established a natural reflex of informing the others about things that could be interesting for them (05/2007).

The partners had established a relationship in which they shared their information and consequently enhanced their understanding of each other, the bureaucratic apparatus and the general scientific and political environment. The advantage of this was that it facilitated rapid reactions to a changing environment in the event that adaptations should be necessary for the stability of the network. The stability of the network couldn't, however, be ensured without the establishment of formal rules capable of structuring the informal interaction processes to a considerable degree. The partners themselves negotiated these rules, preparing a EuroBioBank Charter, a material transfer agreement, a standardised informed consent sheet and a couple of standard operating procedures in which they laid down technical standards for the handling of biomaterials.

3.4 The Evolution of a (Private-) Governance Regime

The regulations governing EuroBioBank contain a mixture of regulatory mechanisms, none of which are statutory laws. An obvious reason for this is that to adopt European trans-national legislation specific to biobanks in the field of orphan disease was regarded as disproportionately complicated and costly, given the tiny number of cases that would fall under that legislation. Further, the use of 'soft law' and private law has the advantage of flexibility and potential for adaptation to an unstable and uncertain environment.

The relationship between the EC and the EuroBioBank consortium, for example, was established through a research grant – a legal instrument from the domain of private law. Relations between donors and individual biobanks are also regulated with the help of the development of a standardised 'informed consent sheet' (IC). This IC was put at the disposal of the biobanking partners without imposing its use on them. Each biobank is free to decide whether it will use its own draft, or the one with standardised content provided by EuroBioBank. Relationships between the partners, including rules for the decision-making process were laid down in the 'EuroBioBank Charter'. The Charter establishes that a representative of EURORDIS is to be the authorized 'administrative coordinator' of the network, and details the rights and responsibilities of all contracting partners. Moreover, the 'Material Transfer Agreement' (MTA) lays down the conditions for the systematic exchange of biological material between biobanks, with a view to stabilising patterns of exchange. The MTA is a binding contract under civil law that governs the relation of legal entities in a policy domain.[4] Finally, the scientists developed a list of 30 Standard Operating Procedures (SOP's) to secure the quality of the samples. The primary requirement of transparency as to the treatment of samples ensures their usefulness in the research context. Despite the fact that they didn't raise any high-tech issues, consensus on the final list of SOP's was difficult to reach, requiring several rounds of negotiation. The problem was that scientists weren't eager to replace established laboratory practices, because, embedded as they were in a complex web of individual biobank procedures, they couldn't be easily changed without a willingness to create a whole new system. Finally an agreement was reached that allowed for the adoption of more than one procedure if there was sufficient assurance that quality standards would be met, even if there was some ambiguity as to whether the alternative procedure was the most efficient solution for promotion of research in the rare disease area.

The IC sheet and the SOP's demonstrate the peculiarity of this governance regime. In the absence of a single actor with the power to determine the goals of the project, it is impossible to judge its efficiency, as the negotiated outcomes are a function of multiple and often contradictory interests. Negotiations take place at the intersection of different, but to some extent overlapping, expert cultures. In this context, it is hard to distinguish between strong arguments and subtle power. The partners are bound to one another, not by an external power, but by their overlapping interests. The weakness, with respect to enforceability, is however (as mentioned earlier) a strength regarding the capability to adapt quickly to an unstable and uncertain environment. The importance of this strength is illustrated in the following section.

---

[4] As a matter of fact these private regulatory mechanisms had to be embedded into the existing legal structures, as well as the ethical practices in which the partners were working already. Therefore a study was conducted on 'outstanding legal and ethical issues' in the emerging area of biobanking and the results were finally published in a booklet. (Uranga et al. 2005) This is an example of the integration of Ethical/Legal and Social Issues Research (ELSI) guiding a research endeavour.

4 The Composition and De-composition of Collective Biosocial Identities

In this section I use an example to illustrate the instability of biosocial identities that are important for stabilizing the relations of EuroBioBank in particular and issues of governance in long term research projects like biobanking in general.

As early as the 1860's the physiologist Guillaume Benjamin Duchenne described the heritable muscular dystrophy that is still named after him today. Duchenne muscular dystrophy (DMD) is a form of muscular dystrophy characterized by decreasing muscle mass and progressive loss of muscle function in male children. Its first symptom is a weakness of the leg musculature appearing between the 3rd and the 5th year. The disease spreads towards shoulders and arms; by age 7 to 12 the boys can hardly lift their arms in a horizontal position and most of them are dependent on a wheelchair. By about age 18 a large proportion of affected persons are totally dependent on care. Only a few live to be older than 30. Characteristic for their physiognomy are the 'pseudo-hypertrophic' calf muscles, which look bulky, because muscle tissue is replaced by fat and connective tissue.

For people suffering from DMD, the advances in genetics were even more promising than for many of the other rare disease patients. In 1986, the dystrophin gene, which is responsible for this devastating disease, was discovered and located on the X chromosome, which meant that for the first time Duchenne patient activists like Barataud, previously mentioned, were provided with a visible enemy. The fight against this 'beast' united the sufferers, and its qualities partly structured their collective actions (Rabinow 1999: 39ff).

Paradoxically, it is the efforts of the patient groups, who through their involvement in research have helped develop an understanding of the DMD condition, that are beginning to construct perceptions of difference within their biosocial group. Potentially it will be their own success that divides them.

On December 3rd an article was published in 'Science' entitled 'Rescue of Dystrophic Muscle Through U7-snRNA- Mediated Exon Skipping' (Goyenvalle et al. 2004). The authors described a technique that could rescue the production of the dystrophin protein, which is responsible for the stability of the muscle cell-membrane and coded on the dystrophin gene. It explains that the cause of DMD symptoms is 'frameshift'mutations: specific mutations on the gene cause an improper reading of the DNA code and finally lead to an unwanted 'stop codon' that terminates altogether the coding procedure that creates dystrophin. These incomplete proteins build only disaggregating muscles, leading to the severe consequences of DMD. The scientists' solution to the problem was to skip one or more of the coding DNA parts (the exons) in order avoid the reading of a 'stop codon'. The resulting – still incomplete – proteins lead to the development of a milder form of the disease called Becker Muscular Dystrophy (BMD).

The interesting point is that the exon-skipping technique is no longer aimed at the whole dystrophin gene. Frameshift-mutations appear at different exons of the dystrophin gene in different patients, who would therefore profit from different skipping techniques. The following quotes are taken from a German internet forum that is dedicated to the communication of Duchenne patients and their families. They illustrate the various perspectives, hopes, and interests that arose within the Duchenne community immediately after the discovery.

> As always the crucial question is, "how long do we have to wait?" "When will the decision be made as to which exons will be skipped after exon 51?
>
> My son suffers from a severe manifestation of BMD! There is a deletion of exons 45-48! Will my son be able to profit from exon skipping as well? For what kind of research do we have to hope?
>
> Sounds nice. However, as you know I am mostly interested in the duplication 6-9 of my nephew, which is not really the focus at the moment. I believe that when one of them works the others are going to follow. I read that there are a lot of these boys out there. These boys will not be forgotten![5] (translation by the author).

The questions and statements started to diverge. The affected people wanted to know when and how decisions would be made, about whose hopes would be considered first. This example illustrates that individual and collective identities comprising the networks cannot be regarded as stable. This is especially relevant in the context of biomedical research and biosocial groups, who have bound their identity to a constantly changing body of knowledge that describes them. Neither is there anything like a uniform, pre-determined patient interest, or a partially existing patient interest, which can be regarded as reliably stable over a certain period of time. Especially in a field like biobanking in which benefits are often depicted as contingent and distant, this could be one of the major governance challenges for the future.[6]

In order to cope with these challenges it will be important to establish models of participation like the ones established in the case of EuroBioBank. Such reflexive organizations "constantly question the procedures and tools enabling it both to learn, i.c. to accumulate competence and knowledge produced collectively, and to evaluate this competence and knowledge so as to decide on future actions to undertake" (Rabeharisoa and Callon 2006: 159). Non-scientific contributions are no longer regarded as obstacles for science but as a specific resource for the whole endeavour. Discussion on topics like the use of biobanks and the development of new protocols should not be left to the discretion of scientific expert knowledge. Appropriately designed biobanks will not only welcome organized patient groups to participate during the construction phase of the project, but will reopen their

---

[5] www.abc-online.org/forum/read.php?f=2&i=6881&t=6881&v=f. Accesed 14 February 08.

[6] This argument can be illustrated by a quote taken from a speech held by John Newton who was appointed CEO of UK Biobank in 2003, that is unfortunately no longer accessible via UK Biobank homepage: "The UK biobank is not a single study or even a single project, but a resource for the biomedical research community for the first few decades of this century. Its true value may not be realized for some 30 years. But every generation has to plant the shade trees for the next."

agenda for discussion on a regular basis, making possible cooperative learning and the building of mutual trust. They are open to respond to new questions, differences, tensions and conflicts in creative and productive ways. They are also able to create new points of interaction for marginalized groups, their voices and their bodies, who want to take an active part in shaping their future health.

5 Conclusion

In this paper I reconstructed the evolution of one of the first operating biobank networks in Europe: EuroBioBank. I argued that we can enrich our understanding of that process if we conceptualize biobanks as a complex socio-political process, instead of a tool or resource that has to be smoothly embedded in a social context after its construction. I demonstrated how genes became a political problem that was eventually tackled on a European level. EuroBioBank developed as an effect of interactive processes between a multiplicity of micro- and macro-actors and the convergence of several policy dimensions, which managed to stabilize each other. I pointed out that the governance structure is one aspect of more comprehensive changes in political decision- and policy-making. In this new governance regime power is fragmented and divided among several players, none of them having the required capacity to assume the leading role. Under these circumstances, rules that regulate conduct and therefore stabilize such inter-organizational systems, are often laid down without the direct involvement of government bodies. After showing that individual and collective identities that comprise a biobank cannot be regarded as stable entities, I recommended that the biobank be constructed as a reflexive organisation that is not directed solely by scientific expert knowledge, and is therefore capable of adapting to a variety of future scenarios in a flexible way. The adoption of private governance regimes that functionally replace statutory law was presented as a promising instrument for transnational biobank network governance.

Acknowledgments   The research on which this article is based was generously funded by the EU's 6th framework Programme for Research project GeneBanC.
I thank Herbert Gottweis, Ingrid Metzler, Heidrun Huber, Ninette Rothmüller and Richard Hindmarsh for their suggestions and remarks regarding some of the issues discussed here. Moreover, I would like to thank all members of the 'Life-Science-Governance Research Platform' for ongoing discussions and the productive research environment they provided. Further I appreciate comments on an early draft by other contributors to this book, especially the editors.

References

Barataud B (1992) Au nom des nos enfants. Paris

Barry A (2001) Political Machines: Governing a Technological Society. The Athlone Press, London/New York

Callon M (2005) Disabled Persons of all Countries, Unite! Making Things Public-Atmospheres of Democracy. In: Latour B and Weibel P, pp 308-313. ZKM/MIT Press, Karlsruhe, Cambridge

Desrosières A (2005) Die Politik der großen Zahlen. Eine Geschichte der statistischen Denkweise. Springer, Berlin/Heidelberg

The European Parliament and the Council of the European Union Decision No.1295/1999/EC (1999). Adopting a programme of Community action on rare diseases within the framwork for action in the field of public health.

Goyenvalle A, Vulin A, Fougerousse F et al (2004) Rescue of Dystrophic Muscle Through U7 snRNA-Mediated Exon Skipping. Science 306(5702): 1796-1799

Keller EF (2000) The century of the gene. Harvard University Press, Cambridge, Massachusetts

Luhmann N (2002) Die Politik der Gesellschaft. Suhrkamp, Frankfurt aM

Mayntz R (1993) Policy Analyse. Kritik und Neuorientierung. Politische Vierteljahresschrift 34: 39-56.

Merkel A (1927) Allgemeines Verwaltungsrecht. Verlag von Julius Springer, Wien/Berlin

Rabeharisoa V, Callon M (2006) Patients and scientists in French muscular dystrophy research. States of Knowledge: The co-production of science and social order. In: S. Jasanoff pp142-160. Routledge, New York 142-160.

Rabinow P (1999) French DNA: Trouble in Purgatory. The University of Chicago Press, Chicago, London

Rabinow P (2004) Anthroplogie der Vernunft: Studien zu Wissenschaft und Lebensführung. Suhrkamp, Frankfurt aM

Stark K, Eder J, Zatloukal K (2007) Achieving k-anonymity in DataMarts used for gene expressions exploitation. J Integr Bioinformatics 4(1): 483-495

Stone D (2001) The policy paradox: The art of political decision making. W.W.Norton & Company, New York

Uranga AM, Arribas CM, d. Donato JH et al (2005) Outstanding legal and ethical issues on biobanks: An overview on the regulations of member states of the EuroBioBank project. Instituto de Salud Carlos II, Madrid

Weber M (1976) Wirtschaft und Gesellschaft. Mohr Siebeck, Tübingen

Weiss R (1994) Learning from Strangers. The Art and Method of Qualitative Interview Studies. New York

Informed Consent and Benefit Sharing in Genetic Research and Biobanking in India

Some Common Impediments in Practice

Prasanna Kumar Patra, Margaret Sleeboom-Faulkner

Abstract   In this paper an attempt is made to understand common impediments in the application of two bioethical principles - informed consent and benefit sharing - in genetic and biobanking research in field situations in India. These evolving principles are discussed and addressed in contemporary national and international bioethical guidelines that reflect the nature of the population and technological systems that they deal with. Notably, the importance of these two principles is that they aim, on the one hand, to protect research participants from exploitation, harm and injustice and, on the other, to impose legal and ethical obligations upon those individuals and institutions conducting research and/or business enterprises. The major considerations that will be addressed by this paper are: firstly, whether there is a true, valid and 'informed' consent procedure that will be practicable, particularly in illiterate, resource-poor and marginalized social settings; and secondly, whether human genetic materials can be considered as resources or property and promoted for benefit sharing arrangements, as has been the case with respect to non-human genetic materials. The study draws on primary information collected in India during December 2006 and May 2007.

1 Introduction

The ethical principles of informed consent and benefit sharing are important, yet contentious, wherever humans are recruited as subjects of biomedical and genomic research. These principles are evolving and are discussed and addressed by contemporary national and international bioethical guidelines that reflect the nature of the population and technological systems that they deal with. Notably, the importance of these two principles is that they aim, on the one hand, to protect research participants from exploitation, harm and injustice and, on the other, to impose legal and ethical obligations upon those individuals and institutions conducting research and/or business enterprises (Hansson 2006, Andada 2005). A major theoretical concern is whether there is a true, valid and 'informed' consent procedure that will be practicable, particularly in illiterate, resource-poor and marginalized social settings. A further question is whether human genetic materials can be considered as resources or property and promoted for benefit sharing arrangements, as has been the case with non-human genetic materials. Some critics argue in favor of abandonment of the traditional form of informed consent, because it is perceived as imperfect due to inherent impediments in its basic elements, and as a result of changes in how people understand genetic knowledge and its implications for them and their communities (Chokshi and Kwiatkowski 2005, Upvall and Hashwani, 2001). In this paper an attempt is made to understand common impediments to the use of these two principles as they are applied in the context of genetic and biobanking research initiatives in India, and to determine whether the principles are still desirable and practicable. This study draws on primary interviews with sample providers, community leaders, researchers and medical doctors based in various communities, hospitals, research institutes and universities in India during December 2006 and May 2007.

1.1 Informed Consent: What Is It All About?

Research guidelines of the Council of International Organizations of Medical Sciences (CIOMS) provide a concise definition of informed consent:

A decision to participate in research made by a competent individual who has received the necessary information; has adequately understood the information, and after considering the information, has arrived at a decision without having been subjected to coercion, undue influence, inducement or intimidation (CIOMS 2002).

The concept of informed consent, with its emphasis on individual autonomy, personal decision-making and the protection of privacy, has been a central component in research ethics since human-rights abuses – including Nazi experimentation on inmates of German concentration camps, and the Tuskegee Syphilis Experiment in which US physicians denied their victims treatment in order to study the course of the disease – resulted in worldwide abhorrence. Informed consent – which requires that research participants be informed of all research uses to which their donated material or information might be put – has accordingly become the gold standard of research ethics (Kegley 2004, Elger and Caplan 2006).

Generally, the principle of informed consent applies to two related, but nevertheless different, settings: medical practice and biomedical research on human tissues and health information. The first of these is about the relationship between doctor and patient. Prior to initiation of a procedure or treatment, a doctor must inform the patient as to details of the treatment, its importance, consequences and risks, and seek his or her consent. In the second setting, that of medical and genetic research, it is mandatory to obtain informed consent before extracting or using the biological material of an individual, whether it be cells, tissue or organs. Only by clearly explaining to the sample provider how the biological material will be extracted and used, and by obtaining his or her consent, can researchers proceed both ethically and legally (Kegley 2004).

The duty of obtaining informed consent is a provision of research ethics, and widely recognized in national and international guidelines as well as legislation (Andada 2005). Although such guidelines do not provide a detailed discussion of the nature and veracity of informed consent to be applied in diverse societal and cultural contexts, a closer look at them suggests that the informed consent debate is centered on respect for donor autonomy and the dissemination of information. The concept of informed consent comprises two ambiguous terms. 'Informed' refers to 'information', raising issues related to the type of information, amount of information, when this information should be given, how such information should be provided, to whom it should be provided and who will provide it. 'Consent' points to the subject participant who may be an individual or a group or a community.

1.2 The Significance of Informed Consent and Related Challenges

The significance of informed consent springs from five basic elements that substantiate its validity in various contexts. The first element is the capacity to consent, by which consent may only be given by a person who is legally and factually capable of doing so. The second is full disclosure of relevant information, which requires the participant to be informed, in language that he or she understands, of the research nature of the study, the purpose of the research, the expected duration of participation and planned follow up, foreseeable risks and benefits, and available and advantageous alternative procedures or courses of treatment. The third element is voluntariness, meaning that the research participation of any individual, group, or community must be voluntary and devoid from force and coercion. The fourth element, adequate comprehension of the information by participants, requires that the prospective participant is competent to comprehend the information. The last element requires that that the participant is given the option to withdraw from participation voluntarily at any stage, without prejudice to the participant (Andada 2005). The right of the subject to withdraw from participation accordingly imposes an obligation on the part of the researcher to refrain from putting undue pressure on the potential subject to participate.

When tested in practical field situations in India, one encounters a gross mismatch between theoretical expectations and the practical implications of these elements: individuals and tribal or caste communities are subjected to genetic intervention programmes as a result of screening programmes and genomics related diversity studies. In most cases, the subject participants are not adequately equipped with information, as research groups fail to fulfill the requirement of providing the relevant information to them.

1.3 Informed Consent in Indian Contexts: Some Common Impediments

Informed consent is an ethical ideal or a moral absolute. The gap between this ideal and the practical situation warrants closer examination. This study asks why and under what conditions the basic elements of informed consent are not addressed or fulfilled. What makes an individual or a group incapable of consenting and how can they become capable to give or refuse their consent? What are the hindrances to achieving these capabilities, and are they evident in different situations? Typical hindrances to the achievement of genuine informed consent include illiteracy, poverty, paternalistic attitudes, cultural barriers, implicit forms of coercion, situational pressure, resistance from health care professionals and a lack of or ineffective regulatory mechanisms. This paper discusses some common impediments or hindrances that were encountered while dealing with informed consent in field situations in India. Considering such limitations, we examine whether there is still a need for informed consent.

1.3.1 Types of Genetic Research

Should informed consent procedures be the same for all types of genetic research? Genetic research encompasses basic research on the human genome, research into the genetic mechanisms involved in disease, research into the genetic basis of behavioral traits, genetic population research, and trials on genetic therapy. Other procedures vary depending on the terms of reference of the research: procedures for data collection, the extent of risk and amount of benefit, the use, implication and application of genetic information. Perhaps the various types of genetic research also have specific requirements and implications for informed consent.

The relevance and importance of informed consent is, for instance, different for epidemiological gene studies into carrier status, genetic diversity and migration history among tribal and caste communities in India than it is for sample collection and studies in genomics and biobanking. Case studies conducted at a local level among participating individuals, community leaders and researchers produced interesting responses. Researchers involved at the community level with genetic interventionist studies such as carrier screening for sickle cell anemia do not always feel the importance of informed consent. The collection of blood samples is simply a means of understanding the carrier status, educating people about the genetic basis of the disease and for providing pre-marital counseling. In the context of tribal and rural communities in India, where most people are illiterate and poor, it is not considered necessary to collect informed consent. As one Project Director said:

Why do we need to ask for people or their community's consent? We are just helping them through screening programme, we are taking their blood, analyzing its, telling them their disease status and then we are working on the data for better health management. We are not storing the sample or supplying it outside. I do not feel there is any need for informed consent in our case (Dr. Dalla, the Project Director of Sickle Cell Project, Chhattisgarh).

In connection with sample collection leading to genomic and pharmacogenomics studies, as well as biobanking, however, researchers recognize the importance of informed consent. Such research involves concern for confidentiality and scope for benefit sharing, hence, a proper informed consent procedure needs to be followed.

1.3.2 Illiteracy

Literacy is universally recognized as the most powerful instrument of social change and the extent of literacy within a population as one of the most important indicators of health and social and cultural development. Illiteracy can undermine informed consent by inhibiting its most basic requirement, the understanding of research details and methods. In the consent-seeking procedure, it is intended that participating individuals and communities enter into a dialogue with researchers who impart basic facts and knowledge as well as the advantages and risks associated with the study for which they are being recruited. The process requires special care to ensure proper understanding and informed decision-making whenever illiterate individuals may be placed at a disadvantage. Indian tribal population groups, which have more chance of being recruited for genomic research as a result of their 'special' genetic make-up, have a dismal 36% literacy rate. Case-studies among certain tribal communities have shown that their participation in genetic epidemiological and genomic research is always manipulated by researchers, who fail to follow a proper informed consent procedure before recruitment. On many occasions, when asked about the kind of information they have received from research teams to whom they have given blood or other genetic samples, tribal and illiterate people fail to answer, due to lack of comprehension. When asked if they have signed any paper before giving their blood, they might reply affirmatively, but fail to explain the content of the forms or what they were intended for. As one tribal man puts it: "Oh! Sir, we are poor, uneducated and unskillful people, what can we understand from that piece of paper? We just followed what the doctor and big people there told us to do". This situation also creates difficulties for researchers, who find it challenging to know how to deal with illiterate people. Further, it is questionable whether it is the role of the researchers to explain both the research procedures and the risk implications it might have for the sample providers.

1.3.3 Poverty

One significant characteristic of tribal and lower caste people in India is poverty. Poverty is not only a question of lack of access to basis amenities of life but the inability to make decisions as to what is good or bad for them. Poverty undermines the spirit of the original meaning of voluntariness in informed consent, in situations where, for example, genetic screening or participation in any kind of genetic research is the only way in which a prospective participant can obtain medical attention. In India it is very common for individuals from marginalized groups to participate in genetic research in order to get medical treatment or medical attention. Case studies in the Sahu caste group from Chhattisgarh state demonstrate that poor living conditions and poverty induce many to attend health camps at their village and then travel to the district hospital, located at some 50 kilometers from their village, in order to receive a health care benefit through a genetic intervention project. People give their blood samples for carrier detection study and then for molecular level study without following proper informed consent procedures. When asked why he had attended the carrier screening health-camp and whether he had signed a consent form, one of them said:

> We were told that if we give our blood, we will be given a free medical check-up and free medicine and if somebody is detected as a sickle cell sufferer, that person will receive free medicine for 21 years. We are poor people, so this was great news for us. Hence we decided to participate and donate our blood, otherwise who else will give us such facilities. We were told that this test will help us in knowing if we suffer from the disease or not and we can stop this disease from spreading in our village and community. We did not sign any form and no consent was taken from me or my family members. I do not know if the camp organizers or doctors have talked to any village leader (Mr. A. Sahu, Chapridi village, translation by the author).

Poverty is an important driving force, pushing people to take decisions to participate in genetic research and to donate biological and health information.

1.3.4 Paternalistic Attitude

In India, doctors, scientists and researchers are held in high esteem, especially by rural and tribal people. Their views, suggestions and instructions are taken for granted and respected. In many instances during the fieldwork, it became clear that the social groups mentioned above understand, interpret and implement the informed consent procedures in a way that differs from the instructions contained in guidelines prepared by the Indian Council of Medical Research (ICMR) and the Department of Biotechnology (DBT). In hospitals, clinics and health-camps, potential donors come into contact with doctors, and they usually leave it up to the doctors to take decisions on their behalf. They believe that when it comes to anything 'medical' the doctor knows better than they do. In such situations it is also difficult for the doctor to stick to informed consent procedures. The need to educate, consult and genuinely inform individual participants or donors prior to their signature of the form of consent creates a dilemma in which physicians have to take decisions on behalf of the patient and their profession. Undoubtedly conflicts of interest at times arise. On one occasion, a medical doctor who was also the project coordinator for a genetic intervention project said:

What's the use of informed consent when many are suffering and dying of diseases, we need to screen them to know their status and take appropriate action for treatment or counseling to ensure that the disease burden is reduced in the society? A mere informed consent will only add to their misery. Do the people who prepared ethical guidelines understand the situation at grass-root level?

The doctor, then, regarded the need to seek informed consent as a legal requirement that impedes research and health care management by consuming substantial resources in the process of information and education about risks and benefits. He also emphasized that it is 'we' the doctors and researchers who know better than the subjects what is good or bad for them in this context.

1.3.5 Cultural Barrier

Culture and social tradition play an important role in the acceptance and feasibility of informed consent procedures at local level. In certain cultures, elderly members, heads of households or heads of the community are meant to take decisions on matters that in other cultures would be considered as a transgression of personal life. Among tribal communities, decisions on health and illness are not a personal prerogative, but rather a matter to be decided upon by the religious or community leaders. Individual decisions bypassing such authorities are considered to be rebellious. Individuals willing to participate in genetic or genomic research as donors or research subjects are expected to obtain prior approval from the community leadership and once the leaders agree to an action, individuals will seldom object to it. Case studies among tribal communities in India show that genetic research investigation groups always arrange community consultation between the local administration, the tribal leadership and the participating subjects before recruiting individuals as subjects for study and sample collection. The job of the researcher will not be successful without the support of the leadership.

The ideal way to conduct the informed consent procedure is a matter of contention. Some argue for oral or written consent and others favor audio and video recording of the whole procedure. Written consent is preferable, but is in some societies treated with suspicion, as a result of their prior experience of exploitation by local landlords and money-lenders. Members of illiterate and marginalized communities have been persuaded, for example, to provide fingerprints, thereby undertaking legal obligations without having been given any information about the transaction they were entering into. As a result, they are hesitant to sign or provide fingerprints as proof of consent when asked by a genomic research group to participate as a donor in a research project.

1.3.6 Ineffective Regulatory Mechanisms

Regulatory mechanisms function to safeguard the interests of researchers, allowing them to further developments in scientific fields, and to protect research participants and their communities from exploitation and injustice. In India, the conduct of biomedical research on human subjects in clinical and community settings is governed by guidelines prepared by two central scientific bodies, the Indian Council of Medical Research (ICMR 2000, 2006) and the Department of Biotechnology (DBT 2002). Theoretically every university, research institute or hospital has its own Institutional Review Board (IRB) that evaluates research projects by thorough examination of the scientific, ethical and legal requirements, as set out by the two scientific bodies. An IRB is intended to be independent of the research group proposing to undertake a research project. Its membership includes a cross-section of academia and society who are qualified to take professional decisions, with a concern for social sustainability. We observed that the decisions of the IRB are not always independent and free from manipulation. They are frequently taken at the behest of powerful project leaders who wish to carry out a research project without properly addressing the ethical and social concerns that will be raised by the prospective research study. If its decision-making is not rigorous and binding on project staff, the IRB gives a free hand to the researcher to manipulate local consent procedures, especially where participants are recruited from illiterate and poor backgrounds. On the other hand, if the IRB is too strict in dealing with ethical concerns and blocks the project from taking off, it may hamper scientific growth.

Another area of concern is the lack of interim monitoring systems. In India the culture of internal IRBs is of recent origin, especially at the level of universities and research institutes. Once the Committee or the Board gives a project the go-ahead, there is no further mechanism or system for evaluation of the project as to whether ethical and social concerns are being properly addressed at field level. There is no system or provision for interim assessment of activities in the field by bringing the research leader to discussion table.

1.3.7 Individual or/and Community

Who is capable of providing consent is a contentious issue in genetic and genomic research. Researchers are divided as to whether to take informed consent from individuals or from communities, or both. A consultation document for the International Development Commission on Intellectual Property commends procedures for obtaining group consent as well as individual consent, but remarks that group consent is not a substitute for individual consent (Dickenson 2004). Although some studies show that collectivist societies do not unconditionally reject individuality (Andada 2005), there is a growing emphasis, on community consent, as cultures in parts of Asia and Africa place little value on personal autonomy, First person consent of individual community members is replaced by the proxy consent of local authorities, leaders and government officials..

Goup consent has been recognized by the Tri-Council of Canada (TCC 1997), Taiwan Bio-bank (Tsai 2006) and by the Human Genome Diversity Project (NARC 1996), in support of the principle that population consent, as well as individual consent, should be sought for genetic research (Godard et al 2003). Many argue in favor of community consent, either in place of individual consent or as supplementary to it. For instance, Willkinson said that one might protect communities through consultation between researchers and communities or its leaders before the project is drafted, during the project period and during the final version of the draft (Willkinson 2004). The question is whether a researcher, having consulted with a community, may ignore its recommendations and proceed with an unmodified proposal. If not, then the community has the power to effectively veto the project. If the community decision is not definitive, then the researchers have the option of proceeding by involving individual participants, but community consent generally seems to restrict individual freedom and may violate the rights of an individual who wants to participate (Willkinson 2004, Godard et al 2003).

In the Indian situation, community consent is a concept borrowed from the traditional anthropological field-work techniques as a means of quickly penetrating the community and obtaining approval to recruit subject participants without antagonizing individuals. It neutralizes somewhat the importance of individual informed consent procedure and fixes the data gathering process (Patra and Sleeboom-Faulkner 2007). As one researcher opined:

If we plan to take some sample from a community, we first approach the village or community headman, the local teacher, local representative or key person who can convince their own people, act as a buffer if there is any resentment, and give some kind of protection to use and help in easy conduct of research. Once they give the permission, getting individual support becomes very easy. It is crucial in community research.

About community consent, a project director who is heading a study on human genome diversity said:

In genetic data generation, we just have informal consultation with the village/community leaders such as village headman, local school teacher or local level Panchayat representatives. Though we give equal importance to individual consent, but, at times with community leaders' approval and consent you can ignore or bypass the individuals.

'Community engagement' and 'community consultation' are ethical bywords used in population-based studies of human genetic variation, where the theoretical aim is to enhance the control of human populations, the research subjects, over the initiation and conduct of the project and the methods by which its members are studied (Juengst 2003). In practice, however, community consent is used strategically for the recruitment of individuals as subjects of genetic research. In the Indian context, any attempt to define a 'community' or a 'group' is problematic, since many population groups share overlapping cultural, social, political, economic and other characteristics. In addition, it is very difficult to identify a 'culturally' or 'politically' appropriate leader with authority over all members of a specified community or group.

1.3.8 Procedural Inconsistency

The documentation of informed consent is not a substitute for the detailed process of procuring consent. The ethical validity of informed consent hinges not on the written word, but on the quality of the interaction between the patient and clinician or researcher and a subject participant; record keeping is just one part of the process. Informed consent is based on mutual trust between investigators and participants. In India, several flaws were observed in the consent-seeking procedure, in that it focused more on documentation than on qualitative improvement in dialogue and trust between researcher and subject. At the community level, it is the researcher who decides which type of procedure best suits his interest, whether oral, written, video or audio documentation. The interests of the individual or group participant are not taken care of. Of course, national guidelines cannot fully address the nature of consent procedures that depend on the specific situation of the community or group from which the informed consent is required. So, it is the researcher who determines the kind or combination of procedures that is best suited to safeguarding the concerns of both the subject participants and the researcher.

Procedures for taking informed consent vary considerably. In some cases oral consent from individuals and/or group are considered to be sufficient, in others oral and written consent are mandatory, and in yet other cases the whole procedure is audio taped or video recorded. It was observed during the fieldwork that procedures varied according to the researchers, in relation to their disciplinary background, institutional affiliations and understanding of ethical principles in biomedical research. Dr Mitashree Mitra, a professor of physical anthropology at PRS Shukla University, Raipur, and director of a project on Human Genome Diversity Studies among four tribal communities in central India, says that it is her long-standing rapport with the community that is most important in obtaining consent. Although she considers written consent as valuable and desirable, she resorts to community consultation and oral consent. Professor PP Majumdar, a prominent figure in genomics research in India and a professor at the Indian Statistical Institute, Kolkata, is of a different opinion. He feels that, "since practicing written consent is tricky in tribal and rural areas, they cannot but resort to a mixture of oral and written consent process for which the entire procedure is recorded and video taped". Dr. Lalji Singh, Director of the Center for Cellular and Molecular Biology (CCMB), Hyderabad, a leading research center in genomics, who claims to have carried out extensive research on tribal communities of Andaman & Nicobar Islands of India said,

> [...] it is a triangular consultation structure between the community, the local tribal welfare board and CCMB through which we seek community consultation and generate support for sample collection. Additionally we also take individual consent.

These case studies demonstrate two basic things. Firstly, researchers do not view the consent-seeking procedure as a method but as a means. Secondly, these procedures are driven largely by administrative requirements, seemingly to protect the researcher from untoward incidents. There is little evidence that researchers attempt to consider the consent-seeking procedure as a process of informing, educating and empowering participants to take 'informed' decisions about their participation.

1.3.9 Blanket Consent

Blanket or 'open' consent is given only once, but covers any use of the collected biological material at any time in the future. This is particularly important for scientific research that aims to use the deposited materials in new projects or experiments years after the consent has been given, when individuals may even have already died (Kegley 2004). The World Health Organization suggested in 1997 that blanket consent might be the most efficient way to facilitate the use of samples in future projects. This suggestion is however open to criticism. One could ask if blanket consent is truly informed as legally required in most jurisdictions. Since future uses are not known at the time consent is given, do we really consider that the participant received enough information to legitimize any future use (Deschenes et al 2001)

Among researchers in India in the field of genomics and DNA banking, blanket consent is the most common consent procedure promoted and practiced. The argument given by researchers in favor of blanket consent is that it avoids logistical and administrative problems. They also feel that truly informed consent covers only specific and known uses of biological material. Even researchers find it difficult to keep sufficiently on top of developments in genomics to be able to provide enough correct information to enable subject participants to give valid consent. Illiterate and resource-poor participants, though, are unaware of the difference between valid consent and blanket consent and the researchers want to play a safe game. Some contend that blanket consent falls far short of truly informed consent, as it is vague and, as such, of little use in legal proceedings. Neither does it allow participants to act on their continuing interest in health information (Caulfield 2002). This viewpoint was shared by the Quebec Network of Applied Genetic Medicine, which stated that "Consent is a continuing process and must be reconfirmed for instance in the cases of significant changes to the research protocol, to the conditions of banking, in the research partnerships, and in the management of the bank" (Cardinal et al 2003).

1.3.10 Retrospective Consent

The issue of retrospective consent is potentially a serious topc of discussion in the context of biobanking and in regard to India's decision to establish a central National Repository (NR). The NR is considering the collection and storage of DNA samples from all anthropologically defined population groups in the country and intends to gather them, both prospectively and retrospectively, from public, private and individual research centers. Recommendations of the National Advisory Committee of the NR, the precise nature and scope of which are not yet public, have been submitted to the central government for approval. If the NR comes into existence, it may need to obtain agreement from donors and sample providers to transferal of their samples, in addition to permission for their future use. To trace providers in order to obtain further consent is a daunting and impractical task, and the fact that blanket consent is the common practice among researchers in India might benefit the NR. A serious ethical concern may arise, however, if donors have a right to know about new arrangements for the use of their samples, with possible implications beyond those discussed at the time of the giving of the blanket consent,.

2 Benefit Sharing

Benefit sharing is one of the most talked about, complex and controversial bioethical concepts in the national and international regulatory frameworks for biomedical research. The term first arose at the Convention on Biological Diversity (CBD), was adapted at the 1992 Earth Summit in Reo de Janeiro. It has since been used to describe the relationship between those granting access to particular genetic resources (human and non-human) and those providing compensation or reward for its utilization (Schroeder and Lasen-Diaz 2006). The Ethics Committee of the Human Genome Organization (HUGO) endorsed the concept and developed it further in the context of human genetic research on April 11, 2000 (HUGO 2000).

Trying to capture its meaning, Schroeder defines benefit sharing as 'the action of giving a portion of the advantage or profits derived from the use of genetic resources or traditional knowledge to the resource providers, in order to achieve justice in exchange'(Schroeder and Lasen-Diaz 2006). Though benefit-sharing has a strong legal sense that cannot be captured by the combined meanings of 'benefit' and 'sharing', the meaning of these constituent parts implies that there is more than one player or stakeholder engaged in an enterprise involving some kind of tangible or intangible profit.

Article 15 of the Universal Declaration on Bioethics and Human Rights published by UNESCO, which focuses on 'benefit sharing', makes two main recommendations. First, "benefits resulting from any scientific research and its applications should be shared with society as a whole and within the international community, in particular with developing countries". Second, "benefit should not constitute improper inducements to participate in research" (UNESCO 2005). The HUGO Statements on Benefit Sharing underpin three fundamental arguments in favor of benefit sharing: human solidarity, the human genome as common heritage and the elimination of exploitation. First, since we share 99.9 % of our genetic make-up with all other humans, we owe each other, in the interest of human solidarity, a share in common goods such as health. Second, like common global resources such as sea, air and space, that are equitably and peacefully available to all humanity, the human genome may be considered a common heritage. Third, benefit sharing can address the concerns of the possible exploitation of less powerful research participants by powerful research organizations (Knoppers 2000).

2.1 What Constitutes Benefit?

In biomedical research, especially in genomics, the definition of 'benefit' is problematic. There exist different categories of benefits including long term and short term benefits, individual and group benefits and appropriate and inappropriate benefits, all dependent on the background of the research participants and the nature of the deal they have with the researchers, research institutes or drug farms. Better health information can itself be considered a benefit, as well as the information process that may lead to therapies and drug deliveries, depending on who it is meant for. Improved understanding of disease processes and the potential for new therapeutic modalities benefit the whole of mankind. Molecular comparisons of healthy and diseased tissue may provide new insights into disease processes and point to avenues for drug discovery that would benefit the pharmaceutical industry and the successful development of new drugs for patients, families and industries (Berg 2001). Short and long term benefits are demonstrated by the provision of free medical services or pharmaceuticals to individuals or populations participating in research, while the development and sale of patented drugs could produce very high revenues for pharmaceutical companies for the duration of the patent period (Berg 2001). Benefits in kind, such as medical help or in any non-monetary form must be appropriate in nature, scale, and distribution within the community. Monetary benefits might raise special concerns about legality and coercion. Transfer of scientific technology is another type of benefit that may be particularly useful in some contexts. The scale of the benefit is also a major concern: an enormous benefit may effectively make it impossible for the community to refuse, thus rendering the process of informed consent meaningless. Finally, researchers are warned against doing harm through a careless distribution of benefits within the community (Godard et al 2003).

2.2 Challenges Involved in Benefit Sharing

The concept of benefit sharing is problematic due to complex stakeholder relations involved in any biomedical research that leads to profit. Significant problems persist even if broad agreement exists on financial benefit sharing with sample-providing individuals or their group. To devise a mechanism that guarantees justice and an equitable share of revenue between stakeholders is a challenge. Imagine a case in which genetic markers from a few individuals or a small community lead to drug development for a deadly or common disease. Who should benefit? Who should not, and on what basis? It is reasonable that key individuals and their relatives must benefit, as it is their 'unique' genetic make-up that leads to a discovery and finally the product that might generate revenue. But others could argue that it was only a matter of chance that these particular key individuals became participants and that anybody with a similar disease profile or belonging to the studied community could have become a beneficiary. Again others might say that there is nothing like a 'unique' genetic makeup; each participating individual has merely inherited some genetically synthesized compounds based on hereditary instructions passed on from parents whom he or she has no control over.

Yet others may argue that individual or community DNA is only 'valuable' as a result of the work researchers have done with it or because of some pre-existing knowledge attained by the work of other scientists at an earlier stage. According to this argument, the people whose work has made a DNA sample valuable must have some stake in it, a right that is closer to that of a holder of intellectual property than any right that the donor of the sample could claim. If industry has a duty to compensate anyone, the scientist's claim to the resulting benefits should exceed that of a person who has merely donated DNA (Berg 2001).

The payment of financial compensation 'upfront' is also problematic for universities and research laboratories that already face financial strain as a result of difficulty recruiting participants. The requirement to pay funds to participants at the beginning of a research project would curtail the growth of research related to drug discovery based on genomic research. Secondly, it is difficult at the outset to assess whether the research will yield a definite result leading to revenue. Numerous attempts at research may lead nowhere, and result in no tangible profit or benefit for anyone, but the concept of benefit sharing is not void in the absence of a benefit. Thirdly, financial compensation of participants in the early stages of a research project may be small, and may prevent the negotiation of a more appropriate share of the revenues that the research may generate in the future.

2.3 Return of Benefit – the Mechanism

Once we identify the various players in the enterprise of biomedical research and recognize the uncertainty of the scope for profit, then it becomes apparent that the issue of benefit sharing is important to all players. Devising a mechanism for benefit sharing is going to be an uphill task. As B.M. Knoppers puts it in her editorial, "Creating specific mechanisms for benefit sharing may well prove difficult, especially in the cases of large groups and multifactorial diseases. Further, profits may accrue many years after the initial research and to a different entity. Patent rights may expire before or soon after a product becomes profitable. It is therefore preferable that companies act directly, voluntarily, and in harmony with community values and preferences" (Knoppers 2000).

In the context of biobanking, individuals generally donate their biological material for altruistic reasons or in order to further the collective good of the community, rather than for their own profit or the profit of a private company (Greely 2001, Godard et al 2003). At the same time, while some policies aim to encourage the commercial exploitation of publicly funded research, there is also a recognition that this must be matched by a suitable social return, either in the form of technology transfer, local training, joint ventures, provision of health care or information infrastructures, reimbursement of costs, or the use of some portion of royalties for humanitarian purposes (HGO 2000, ICMR 2000, Godard et al 2003).

Though there exists no unanimity among national and international guidelines regarding the modalities of benefit sharing, the provision of benefits to certain participants and populations has been proposed in many jurisdictions. According to the North American Regional Committee of the Human Genome Diversity Project three basic principles should govern researchers with respect to benefit sharing. First, honesty: the researcher must inform the community as to what may be expected from the research and warn them that the research results may not yield practical benefits. Any benefit that is promised must be both deliverable and delivered. Second, legality: national and local laws must be consulted to determine the sorts of benefits that may legally be given to whom (Godard et al 2003). Third, appropriateness:, the benefits must be appropriate in their nature, in their scale, and in their distribution within the community.

In India, where informed consent is not properly obtained and the identity of individuals is not recorded or maintained, a fair and just mechanism for benefit sharing is highly unlikely to be secured. Though benefit sharing is discussed in existing protocols (ICMR 2000 and DBT 2002), the mode and mechanism of sharing between unequal players (powerful authority and less powerful subjects) is not addressed fully. The DBT (2002) guidelines, for instance, say, "it is obligatory for national/international profit making entities to dedicate a percentage (e.g. 1% - 3%) of their annual net profit arising out of the knowledge derived by use of the human genetic material, for the benefits of the community."

2.4 Benefit Sharing Mechanism in India: the Case of Kani Tribe

"India has pioneered one of the first models of benefit sharing" (Iype 2002). This statement was given by R.A. Mashelkar, Director General of the Council of Scientific and Industrial Research (CSIR), hailing the successful experiment of a benefit sharing model called 'Kani Model' or 'Pushpagadan's Model' (NBRI 2002).

The Kani tribe, with a 1991 population of 16,181, is a traditionally nomadic but now settled community, barely making a living in the forests of the Agasthyamalai hills of the Western Ghats region of Kerala State in India. Like many other indigenous communities, the Kani use plants for a variety of therapeutic as well as nutritional purposes. In 1987, a government initiative, the All India Coordinated Project on Ethnobiology (AICRPE), sent a team of scientists from the Tropical Botanic Garden and Research Institute (TBGRI)to undertake an ethno-botanical field study in the area of tribal habitation. During this expedition, it was discovered that the Kani people were using Trichopus zeylanicus travancoricus, a rhizomatous herb with anti-fatigue properties. When, during a field trip, TBGRI scientists were offered the leaves of the plant, they noticed 'a sudden flush of energy and strength' (Schuklenk and Kleinsmidt 2006, Pushpagadan 1988). A TBGRI analysis of the leaves of the plant later 'revealed the presence of certain glycolipids and non-steroidal compounds which have anti-stress, anti-hepatotoxic and immunomodulatory / immunorestorative properties' (Anuradha 1998).

After satisfactory phase 1 to 3 clinical trials, the herbal drug was renamed 'Jeevni' for commercial production. Following WHO guidelines, therapeutic efficacy, safety and shelf life were verified and standards applied in regard to the manufacture of the drug. Negotiations resulted in the transfer of the licence to manufacture 'Jeevni' to the Aryavaidya Pharmacy Coimbatore Ltd for a fee of Indian Rupees 10 Lakhs. The licensing provisions included a term of 7 years and royalties of 2.0% of ex-factory sale prices. In November 1995, the CSIR norms were adopted in regard to the transfer of technology. In consultation with the tribal community, the TBGRI was able to work out an arrangement for benefit sharing, in which 50% of the license fee and royalties will go to the tribal community through the Kerala Kani Samudaya Kshema Trust. The Trust was founded in 1997 and received its first payment under the benefit sharing agreement in 1999 (Schroeder and Lasen-Diaz 2006).

Benefit sharing aims to achieve an equitable balance between the provision of access to a genetic resource and the payment of compensation. The Convention on Biological Diversity (CBD), adopted at the 1992 Earth Summit in Rio de Janerio, is the only international legal instrument that sets out obligations for sharing benefits derived from the use of biodiversity, but it excludes from its scope human genetic resources (Schroeder and Lasen-Diaz 2006).

The interesting question is whether the model for sharing benefits associated with non-human genetic resources, as in the Kani tribe, can be applied to human genetic resources. Some argue against the use of this model, on the basis of the essential differences between human and non-human genetic resources (Schroeder and Lasen-Diaz 2006) and give three main reasons for their opposition. First, they suggest that it might make donors of DNA more vulnerable to exploitation by negotiation of agreements that will restrict benefits to future improvements in health. Secondly, the potential for inappropriate inducement is considered to be higher in the context of human tissues than in relation to traditional knowledge, and benefit sharing is not a panacea for shifting government responsibilities. The third view is that the application of the arrangement for benefit sharing in human genetic resources will encourage the commodification of human body parts or genetic material, a practice that many believe to be morally corrupt.

3 Conclusion

This paper has outlined a variety of factors that might impede the implementation of informed consent in field situations in developing countries such as India. Some of the factors are intrinsic to the socio-cultural life of those recruited as research subjects, while others are related to the introduction of new genetic knowledge and technologies.

Illiteracy and poverty should not be used as a justification for the omission of valid informed consent. In India, it is people from this segment of society who are most likely to become the subject of study. Paternalistic attitudes in particular hinder the practice of informed consent. As shown, they impose a dilemma for the subject as well as the doctors and researchers. Subjects feel incapable of taking 'informed' decisions on matters of importance and rely on doctors and researchers to take decisions on their behalf. This reliance should not be taken as a pretext for non-compliance with the requirements of informed consent, even though in some cultural groups informed consent may contravene normal social and cultural conventions of decision-making. Doctors and researchers may feel compelled to take decisions on behalf of the subjects, or other situations may feel unobliged to seek subjects' views or consent on matters of genetic sampling.

There has been a tendency to replace true and valid consent by the application of various other procedural means and methods. Methods such as oral, written and audio-video recordings are however mere documentation of the consent procedures. Effort has to be made to educate, inform and empower people to take 'informed' decisions on their own behalf.

The role and importance of regulatory bodies such as the Institutional Review Boards in genetic and genomics research cannot be overemphasized. An independent, unbiased and professionally driven Board is crucial. The issue of individual or/and group consent is not limited to the determination as to who is legally competent to give consent, but also addresses the identification of the representatives of a social group or a community.

Blanket consent is a procedure used by some researchers to minimize the ability of participants to exercise influence over the future use of their samples and materials. Blanket consent should be discouraged in developing countries, such as India, where it is commonly practiced and promoted, but where regulatory mechanisms are ineffective. Some feel that blanket consent falls far short of true informed consent, as it does not allow participants to act on their continuing interest in health information (Caulfield 2002, Kegley 2004).

Finally, although there is broad agreement at national and international levels on the benefit of informed consent as an essential bioethical tool, various jurisdictions such as India are not conducive to its application. What then needs to be done? Appropriate social, political, legal and scientific conditions are obviously necessary to empower each individual donor or subject, as well as his community, to take decisions that are beneficial to them.

Discussions regarding the potential for benefit sharing in genomics and bio-banking research are often based on suppositions, rendering any conclusions on their application presumptuous. Firstly, what constitutes benefit, in diverse social and ecological contexts, is problematic. Better health care or enhanced knowledge of a disease may be considered as a form of benefit to some, while financial profit may be the only valid source of benefit to others. Secondly, it is almost impossible to identify rightful stakeholders and devise a meaningful benefit sharing mechanism based on equitable distribution. A classic example of this is the situation of the Kani community, as discussed in this chapter. There remains, however, a big question mark as to whether this arrangement for benefit sharing in non-human genetic resources may be applied to human genetic resources.

Apart from the moral debate over commodification of body parts, and the legal challenge over rightful stakeholders, benefit sharing can be considered as a means of recognising and empowering subject participants as partners. The benefit sharing mechanism could be adapted to diverse social contexts, in accordance with the local social values and legal frameworks, while protecting against undue inducement.

Acknowledgments   The research for this article was conducted within the framework of the research project Socio-genetic Marginalization in Asia Programme (SMAP) financed by The Netherlands Organization for Scientific Research (NWO) at International Institute for Asian Studies, Leiden University, The Netherlands. The authors would like to thank all the respondents in the field who participated in the research.

References

Andada P (2005) Module two: Informed consent. Dev World Bioeth 5 (1): 14-29
Anuradha RV (1998) Sharing with the Kanis: a case study from Kerala, India. Kalpavriksha Mimeo, New Delhi.
Berg K (2001) The ethics of benefit sharing. Clin Genet (2001)59: 240-243
Cardinal G, Deschenes M, Knoppers BM et al (2003) Statement of principles on ethical conduct of human genetic research involving populations.
Caulfield T (2002) Gene banks and blanket consent. Nat Rev Genet 3: 577.

Chokshi DA, Kwaiatkowski DP (2005) Ethical challenges in genomic epidemiology in developing countries. Genomics Soc Pol l.1(1): 1-15

CIOMS (The Council of International Organizations of Medical Sciences) (2002) International Ethical Guidelines for Biomedical Research Involving Human Subjects: guidelines 4,5 and 6

DBT (Department of Biotechnology, Government of India) (2002) Ethical Policies on Human Genome, Genetic Research and Services. http://dbtindia.nic.in/policy/polimain.html Accessed on 12 February 2007

Deschenes M, Cardinal G, Knoppers BM et al (2001) Human genetic research, DNA banking and consent: a question of 'form'? Clin Genet 59: 221-239

Dickenson D (2004) Consent, commodification and benefit sharing in genetic research. Dev World Bioeth 4 (2): 109-124

Elger BS, Caplan AL (2006) Consent and anonymization in research involving biobanks. Eur Mol Biol Organ Rep 7 (7): 661-666

Godard B, Schmidtke J, Cassiman JJ et al (2003) Data storage and DNA banking for biomedical research: informed consent, confidentiality, equality issues, ownership, return of benefits. A professional perspective. Eur J Hum Genets 11 (2): 88-S122

Greely HT (2001) Informed consent and other ethical issues in human population genetics. Ann Rev Genet 35: 785-800

Hansson MG (2006) Building on relationships of trust in biobank research. J Med Ethics 31: 415-418

HUGO (Humane Genome Organization) Ethics Committee (2000). Statement on Benefit Sharing. http://www.gene.ucl.ac.uk/hugo/benefit.html Accessed on 6 January 2006

HUGO (Human Genome Organization) (2000) Statement on Benefit Sharing. Vancouver, 9 April. www.hugo-international.org/ statement_on_Benefit_Sharing.htm Accessed on 6 January 2006

ICMR (Indian Council of Medical Research) (2000) Ethical Guidelines for Biomedical Research on Human Subjects. ICMR Bulletin 30:10:107-116.

ICMR (Indian Council of Medical Research) (2006) Ethical Guidelines for Biomedical Research on Human Participants. Published by Director-General, Indian Council of Medical Research, New Delhi.

Iype G (2002) And benefits for all. The Rediff Special News October 11. www.rediffmail.com/news/2002/oct/11spec.htm Accessed 16 March 2007

Juengst ET (2003) Community Engagement in Genetic Research: The "Slow Code" of Research Ethics? In: Knoppers BM (ed) Populations and Genetics – Legal and Socio-Ethical Perspectives. Martinus Nijhoff Publishers, Leiden/Boston

Kegley J (2004) Challenges to informed consent. Challenges to informed consent. Eur Mol Biol Organ Rep 5 (9): 832-836

Knoppers BM (2000) Population genetics and benefit sharing. Com Genet 3: 212-214

Knoppers BM (2000) Genetic benefit sharing (editorial). Science 290: 5489, 49

NARC (North American Regional Committee) (1996) Human Genome Diversity Project. Houst Law Rev 33: 1431-1473

NBRI (National Botanical Research Institute) (2002) P. Pushpagandan Model of benefit sharing. www.nbri-lko.org/ director %20data /index10.htm Accessed on 7 April 2006

NCB (Nuffield Council on Bioethics) (2002) The ethics of research related to health care in developing countries. London

Patra PK, Sleeboom-Faulkner M (2007) Genetic biobanking in India – a community based perspective on ways and means of data generation. Taiwan J Law Technol Pol 4:67-97

Pushpagadan P (1988) 'Arogyappacha' (Trichopus zeylanicus): The 'Ginseng' of Kani tribes of Agasthyar Hills (Kerala) for Evergreen Health and Vitality. Anc Sci Life 8: 13-16

Schroeder D, Lasen-Diaz C (2006) Sharing the benefits of genetic resources: from biodiversity to human genetics. Dev World Bioeth 6(3): 135-143

Schuklenk U, Kleinsmidt A (2006) North-South benefit sharing arrangements in bioprospecting and genetic research: a critical ethical and legal analysis. Dev World Bioeth 6(3): 122-134

Tri-Council of Canada (1997) Medical Research Council, Natural Sciences and Engineering Re-search Council: Social Sciences and Humanities Research Council, Code of Conduct for Re-search Involving Humans. Ministry of Supply and Services Canada, Ottawa

Tsai D (2006) Mediating media effects within public participatory opinion in structuring Taiwan Bio-bank. Paper presented at 2006 ELSI Symposium on Reexamining the ELSI Implication of Biobanking – A Cross Culture Perception held at Taipei 17-18 September

United Nations Educational, Scientific and Cultural Organization (UNESCO) (2005) Universal Declaration on Bioethics and Human Rights. Article 15(1). http://portal.unesco.org/shs/en/ev.php.URL_ID=1372&URL_DO=DO_TOPIC&URL_SELECTION=201.html Accessed on 24 May 2007

Upvall M and S Hashwani (2001). Negotiating the informed-consent process in developing countries: a comparison of Swaziland and Pakistan.48 (3) pp. 188-192.

Willkinson, T.M. (2004) Individualism and Ethics of research on Humans. HEC Forum16 (1): 6-26

WMO (World Medical Organization) (1996) Declaration of Helsinki.BMJ; 313 (7070): 1448-1449

Benefit-sharing, Biobanks and Vulnerable Populations

Agomoni Ganguli-Mitra

Abstract   With the rapid development of tools for DNA extraction, there has been a notable increase in the establishment of small and large biobanks for the purpose of research. Combined with highly efficient methods for DNA analysis, biobanks provide a large scale tool for biomedical research. As biobanks continue to be established worldwide, new and recurring ethical issues are put forward in the international arena. One such topic is benefit-sharing, which offers ways to address, on a small scale, existing global inequities, especially those related to the benefits and burdens of medical research. This paper attempts to address some of the concerns related to the development of benefit-sharing frameworks, particularly in the context of vulnerable populations, who typically suffer from the inequities that benefit-sharing intends to address. The discussion is based implicitly and explicitly on the following five questions: Who are the vulnerable? How can benefit-sharing be justified? What exactly should be shared? When is benefit-sharing appropriate? Why are certain benefit-sharing schemes being offered?

Emphasis is put on the importance of the context in which benefits are shared. Furthermore, obstacles and complexities of benefit-sharing are illustrated with regard to the findings of a qualitative study. The paper concludes by looking at biobanks in connection with global justice in the field of health research.

1 Setting the Scene

Since the completion of the Human Genome Project, genomics and genetic medicine have been increasingly perceived as promising areas of biomedical research. Biobanks[1], which provide large scale, long-term storage of biological samples, together with relatively inexpensive methods of genetic analysis, have become prime tools for such research. Some commentators suggest that genomics brings "the ultimate promise of revolutionizing the diagnosis and treatment of many illnesses" (Collins and McKusick 2001: 475). Moreover, according to a WHO re-

---

[1] While the term 'biobank' is defined in various ways in the literature, the following general definition is used for the purpose of this paper: biobanks are collections of human tissues, blood, cells or DNA that can be used for genomic research and that can be associated with personal data.

port, genomics also holds "considerable potential for improving the health of the developing countries in the future" (WHO 2002: 105). As biobanks continue to be established worldwide, new and recurring ethical issues are put forward in the international arena. One such topic is benefit-sharing, which offers ways to address, on a small scale, existing global inequities, especially those related to the benefits and burdens of medical research. This paper attempts to address some of the concerns related to the development of benefit-sharing frameworks, particularly in the context of vulnerable populations, who typically suffer from the inequities that benefit-sharing intends to address. The discussion is based implicitly and explicitly on the following five questions: why, when, who, what and how? (see also Ganguli-Mitra 2008).

1.1 Who are the Vulnerable?

Although 'vulnerability' is a widely used term in research ethics, there is little consensus on how to define it. The debate is taking place in a multitude of different contexts and it is often difficult, if not undesirable, to apply pre-defined characteristics of vulnerability universally. In one of its most extensively used guidelines, the Council for International Organizations for Medical Sciences (CIOMS) uses the following definition: "Vulnerable persons are those who are relatively (or absolutely) incapable of protecting their own interests. More formally they may have insufficient power, intelligence, education, resources, strength, or other needed attributes to protect their own interests" (CIOMS 2002, § 13). In ethical guidelines, vulnerable persons include children, the elderly, those incapable of giving consent, hierarchical subordinates, minority groups (CIOMS 2002) and those economically and socially vulnerable (NBAC 2001: 90). Nevertheless, it must be noted that not all persons who fall under the category of, for example, the economically disadvantaged, are systematically vulnerable (Denny and Grady 2007: 383-384). A thorough analysis of the concept of vulnerability is, however, beyond the scope of this paper. Vulnerability is used here as an umbrella concept to refer to those persons, groups or populations who may, as a result of various attributes or circumstances, be susceptible to harm and exploitation in the context of biomedical research. The term can also act as an indicator of an important power imbalance that might sometimes exist between participants and those conducting research.

1.2 The Special Case of Biobanks and Genomic Research

With the rapid development of tools for DNA extraction, there has been a notable increase in the establishment of small and large biobanks for the purpose of research. Combined with highly efficient methods for DNA analysis, biobanks provide a large scale, yet cost-effective (Collins 2004: 475) tool for biomedical research. Today, biobanks come, so to say, in all shapes and flavours: small, medium, large, international, local, population-based, disease-based; some contain only tissues, others only DNA extracts, some only information and others a combination of all three.

Biobanks, like genomic research, share many ethical issues with clinical research, but also present their own. As such, biobanks in their varied forms might have different impacts on the discussion on vulnerable populations as well as on possible justifications for benefit-sharing. The commonly encountered discourse on risk, for example, is set in the context of clinical research. In genomic and biobank research, the risk discourse takes a different turn as interventions are not necessarily intrusive and physical harm is considered minimal. The risks of stigmatization and discrimination are real but these are associated with use (and misuse) at a later stage in research, or following the dissemination of results, and not typically associated with the intervention and collection of samples. A further point, with significant impact on benefit-sharing frameworks is that, due to the shared nature of DNA, the delineation of groups, communities and populations becomes fuzzy, let alone the concept of vulnerable populations. In fact "genetic research as undertaken in the form of population biobanks swells the number of participants considerably and also blurs the very concept of participant" (Simm 2007: 163). Finally, in biobank-based research, specific research goals are not (and cannot) always be clearly established when samples are collected. Given the wide base and the open-ended nature of such research, benefits arising from research may become contingent and distant.

2 Benefit-sharing: at a Crossroad of Justifications?

Apart from a few scholars who have argued for establishing participation in biomedical research as some form of moral duty (Harris 2005), it is generally accepted that participation is based on altruism and solidarity. However, if participation is discussed under the risk-benefit discourse, that is, if the risk for participants must be weighed against the potential benefits research might offer, then the reasons to participate in biobank research may seem even more compelling, since the risks or burdens are considered minimal. Moreover, biobanks are often referred to as global public goods (Knoppers and Fecteau 2003) and if we all benefit from such enterprises, a case could perhaps be made for a duty, however weak, to contribute.[2]

In the face of current global inequities, however, it seems highly unlikely that such an approach will or should be extended to the international context. With the increase in the commercialization of biomedical research, in which research participants are generally less advantaged (vulnerable) and benefits and profits are enjoyed by those who are relatively much better off, it seems unfair to impose a universal duty to participate in research. It is worth noting as an aside here that a similar argument could be extended to minorities or otherwise potentially vulnerable groups within large well-off populations. Although this may warrant a more detailed analysis, questions of solidarity, altruism and benefits may play out significantly differently even within a relatively contained group.

It may not seem an obvious step, in this altruistic setting, to derive a systematic obligation for researchers to share the benefits of research with participants. Increasingly, however, concerns for global justice and fairness have fed the principle of benefit-sharing in the international context. Populations and individuals who already suffer from inequity should not be made to bear the burden of research, unless they are recognized in some way for their participation.

Having said that, it is also important to note that benefit-sharing in the context of genomic research is, one might say, at a crossroad of justifications. Two other bases for justification have been offered by commentators in this discussion, although both are related to, and are often conflated with, considerations of global justice. One of the earliest agreements on benefit-sharing is the 1992 Rio Convention on Biological Diversity, aimed at limiting biopiracy and the exploitation of indigenous biological resources. The benefit-sharing discourse regarding human genomic research draws some of its strength from this model. Nevertheless, given the controversial nature of the debate on owning body parts and bodily material, it is difficult, if not undesirable, to extend such a model to human genetic material. The other potential basis of justification is the medical model in which the risk associated with a contribution to research calls for a return based on the concept of reciprocity (Simm 2007: 163). Such a model is, however, usually restricted to the therapeutic benefits that result from research. Given the specific case of biobanks (especially regarding risk and the contingency of benefits), as discussed above, this model is unlikely to fully justify the sharing of financial profits or intellectual property rights.

While the various models together provide a strong justificatory background for benefit-sharing in biobanks, they might have different roles and weight in the various frameworks of benefit-sharing that are developed and on the type of benefit that is deemed appropriate.

---

[2] See also the article of Peter Dabrock within this publication.

3 Sharing What?

Having looked at the various arguments in favour of benefit-sharing, the following question now arises: what is considered a benefit, and which benefits are to be shared? Existing benefit-sharing frameworks include, among others, therapeutic benefits (where available), shared intellectual property rights (although this is more popular in the case of non-human genetic material), and research capacity building for the participating community (Schulz-Baldes et al 2007: 8).

Two of the most prominent sets of international guidelines on human genetic material review the type of benefits that might arise from genomic research. The Statement on Benefit-Sharing (2000) of the Human Genetic Organization (HUGO) discusses at length the difficult concepts of community, common heritage, justice and solidarity, all of which are crucial to HUGO's vision of benefit-sharing. Benefits range from results of research and health benefits to the sharing of profits generated by profit-making entities for disbursement towards health care infrastructure or humanitarian efforts. UNESCO's International Declaration on Human Genetic Data (2003) also takes a general approach in Art. 19, recommending that "... benefits resulting from the use of human genetic data ... should be shared with the society as a whole and the international community." Like the HUGO statement, the UNESCO declaration lists benefits possible within frameworks of domestic laws and international agreements. These include: special assistance to persons or groups that have taken part, access to medical care, provision of new diagnostics, facilities for new treatments or drugs stemming from the research, support for health services and capacity building facilities. A few procedural questions come to mind: who is to decide which benefits are fair, under which framework should they be shared and who is entitled to benefit?

4 Benefit-sharing in the Real World

4.1 Obstacles and Complexities: Input from a Qualitative Study

In a qualitative study entitled Human Genetic Databases: Towards a Global Ethical Framework, researchers probed issues related to benefit-sharing in the context of biobanking and genomics.[3] Interviews were conducted with 88 international experts, all of whom had been associated with scientific and/or ethical and regulatory aspects of biobanks. Benefit-sharing was addressed in the context of a fictional scenario in which researchers approached the representative body of an indigenous community for permission to recruit the members of the community into a pharmacogenetic study (Elger et al 2008). In this scenario, various benefit-

---

[3] For an extensive report of the results see Ganguli-Mitra 2008.

sharing schemes were discussed: the provision of free genetic tests arising from research, the donation of medical equipment, the sharing of profits and the ownership of intellectual property rights. Intellectual property rights were not offered by the researchers but demanded by the representatives.

The results of the study suggest that the spirit of reciprocity and considerations of global (distributive) justice are fundamental to benefit-sharing. Although no pattern or significant preference for one or more options could be detected, all respondents were in favour of some form of benefit-sharing. Some respondents felt that benefit-sharing was only truly represented by the provision of genetic tests arising directly from research, whereas a share of financial profits was, by contrast, unrelated to the research and potentially exploitative. Others felt that it was the contingent nature of the availability of genetic tests or of financial profits that was characteristic of benefit-sharing, whereas a promise to donate medical equipment early in the study would lead to undue inducement. As expected, concerns were raised on a number of fronts, particularly on various procedural aspects of benefit-sharing. Among the most notable were concerns regarding various forms of exploitation, including those that might be perpetrated by the so-called representatives of the community, and the importance of contextualizing benefit-sharing (Ganguli-Mitra 2008). The next section looks at these two concerns as they are pivotal to further investigation into the necessary conditions and obstacles to be overcome in the furtherance of benefit-sharing endeavours.

4.2 Benefit-sharing: Conditions and Further Concerns

All ethical principles must be set within a procedural context in which other 'ethical' pillars have been rigorously strengthened. This has been reiterated in various ways in the ethics literature, but it perhaps bears repeating. Certain fundamental conditions must be met if benefit-sharing is to address the claims of justice in a way that it does not become a source of harm and exploitation itself.

4.2.1 Revisiting 'When', 'Who' and 'How': the Importance of Context

Context-specificity is particularly important if we are to establish efficient and fair benefit-sharing frameworks. The following questions, for example, must be addressed: when does the requirement for benefit-sharing apply? Is it only when biobanks are set up for a particular disease, or is it even in effect in the case of genetic add-ons, that is, when additional samples are donated for future, unspecified and potential use (Pullman and Latus 2003)?

In the context of biobanks, to whom are benefits due? To participants only or to their community as a whole? Should benefits accrue only to populations that are considered vulnerable and disadvantaged? Is it possible to delineate a community for the purpose of research? Is it possible or at all desirable to artificially delineate a community in terms of shared genes? Might such artificial delineation have deleterious effects on traditional community affiliation (Juengst 1998: 192)? Understanding the local culture and tradition is important for the development of appropriate frameworks, especially in a study of international scope. As the results of the above study have shown, however, it is not at all clear that requirements such as prior consultation with the community (see HUGO guidelines) will solve issues such as exploitation by local authorities.

The importance of context is of course relevant not only to the recipient(s) of benefits but also to those who will share benefits. As mentioned earlier in this paper, the question is who should be obliged to share benefits, under which circumstances, and to what extent. While stringent regulations may require for-profit companies to share the benefits and profits that accrue from research, obligations may also prove to have negative effects over time by acting as a disincentive for non-profit actors, especially if such a scheme is not contingent on profits being made (Berg 2001: 241). Also related to the obligation to share benefits is the follow-up of benefit-sharing schemes. If there is no provision for consistent, long-term ethical review of research, how are we to ensure that benefits are in reality being shared?

4.2.2 Vulnerable Populations and the Problems of Consent and Power

Closely entwined with benefit-sharing is the procedure of informed consent. Consent is in itself an immense topic and far beyond the scope of this paper, which seeks to highlight the relationship between consent and benefit-sharing only. It is accepted that no research participant gives 'fully' informed consent and the average education in an industrialized country may not ensure a clear understanding of what it means to participate in genomic research. Nevertheless, consent procedures are necessary, and must be carried out carefully in relation to vulnerable participants, for whom a lack of education or a different cultural understanding of genes and heredity might prove an obstacle to adequate appreciation of what is at stake. This might be particularly important if such participants are familiar with the risks and discomfort of traditional clinical research, and do not fully comprehend the burdens of research typified by low levels of discomfort. In a population in which economic resources are scarce, benefit-sharing may easily become a source of undue inducement, especially if negotiated at an early stage, when consent is sought. Undue inducement and other issues related to seeking consent for samples, as well as problems related to the commodification of body parts have been eloquently articulated in Donna Dickenson's much-cited article on this topic (Dickenson 2004). Dickenson promotes enhanced and refined consent procedures as the strongest tool of protection for vulnerable populations in the context of benefit-sharing.

In the context of research (as in many other contexts) vulnerability is often expressed as a difference in power: one party is less powerful and therefore more vulnerable in relation to the other. In the context of benefit-sharing, this presents us with various difficulties. Given that there is no systematic way of imposing benefit-sharing schemes on researchers or sponsors, the development of such frameworks may simply boil down to negotiations between the two parties. Since vulnerabilities exist to various extents, negotiations might result in varying degrees of access to benefits, with those most vulnerable and most in need receiving the least. If benefit-sharing becomes a matter of contract, group size and other characteristics of research participants will have an important effect on the benefits being shared. A patient group in a European democracy may have much more bargaining power than the government of a poor nation. Governments of large developing countries will have greater power than those of a small one ravaged by war or corruption. In population databases, the voices of minorities may be drowned in the mass, even within a democracy, despite being strongly represented for scientific purposes.

Shifting responsibility onto a representative body may also prove to be detrimental. In the study mentioned above, some respondents ruled out the option of transferring intellectual property rights to community representatives, stating that this would place too much power in their hands. While the use of a public authority may not be a bad thing if the governing council truly is a representative of its people, otherwise it would have damaging effects, particularly if its position is used to oppress the most vulnerable of its members. If attractive benefits were to be offered by a researcher to a governing body in such circumstances, one wrong may be replaced by another, that is, in trying to avoid exploitation on one level, they might succeed in encouraging it on another level, in relation to vulnerable individuals.

5 Final Thoughts

In reviewing research protocols and benefit-sharing schemes, an ethics review committee might want to ask the following questions: why, in this context, is a benefit-sharing scheme being offered,[4] and why in particular this scheme rather than another? For example, does contribution to the development of a local hospital help to fill a gap in the local or national provision of health? If so, which one? As some commentators have pointed out, benefit-sharing should not replace the work of a government, where it has the resources to address such needs (Schüklenk and Kleinschmidt 2006: 122). If it becomes a political tool, benefit-

---

[4] I am grateful to Prof. Nikola Biller-Andorno for encouraging me to probe further, beyond the theoretical justifications of benefit-sharing, into practical aspects as to *why* benefits should be shared in the first place, and into which inequities it should or should not be expected to redress.

sharing will have failed miserably to address the needs of vulnerable persons and groups. It would indeed be an appalling situation if, for example, benefit-sharing were to become a source of income for corrupt health ministries to be used to strengthen other means of oppression. While this is a rather pessimistic outlook, such concerns must be addressed if the benefit-sharing frameworks developed are to meet the demands of justice.

One final concern remains: of all the possible uses of benefit-sharing, it should not become another excuse for the medical industry to persist in the current global research portfolio. The 10/90 gap (GFHR 2002) is a frequently cited and deplorable statistic but one that must be kept in mind in any ethical debate regarding biomedical research. While the WHO report mentioned in the introduction (WHO 2002) places high expectations on genomic research for the health of developing countries, these expectations will certainly not be fulfilled if the current research agenda persists. The minimal risks and discomfort associated with biobank research does not mean that it should take precedence over more urgent health needs. Biobanks should not be established, with the help of those less well-off, only to find cures for diseases prevalent in industrial countries. If global justice is to be taken seriously, a positive change in the health research agenda should probably be the first point on any of our benefit-sharing lists.

Acknowledgments I am grateful to Prof. Christoph Rehmann-Sutter, Prof. Nikola Biller-Andorno and Dr. Roberto Andorno for helpful comments and guidance in the preparation of this chapter.

References

Benatar S (2007) New perspectives on international research ethics. In: Häyry M, Takala T, Herissone-Kelly P (eds) Ethics in Biomedical Research. Rodopi, Amsterdam/New York

Berg K (2001) The ethics of benefit-sharing. Clin Gen 59: 240-243

Collins FS (2004) The case for a US prospective cohort study of genes and environment. Nature 429: 475-477

Collins FS, McKusick VA (2001) Implications of the Human Genome Project for Medical Science. JAMA 285: 540-544

Council for International Organizations of Medical Sciences [CIOMS] (2002) International Ethical Guidelines for Biomedical Research Involving Human Subjects. www.cioms.ch/frame_guidelines_nov_2002.htm. Accessed 10 September 2007

Denny CC, Grady C (2007) Clinical research with economically disadvantaged populations. J Med Ethics 33: 382-385

Dickenson D (2004) Consent, commodification and benefit-sharing in genetic research. Dev World Bioeth 4: 109-123

Elger B, Biller-Andorno N, Mauron A, Capron AM (2008) Ethical and Regulatory Aspects of Human Genetic Databases. Ashgate, Aldershot (in press)

Ganguli-Mitra A (2008) Benefit-sharing and remuneration. In: Elger B et al (eds) Ethical and Regulatory Aspects of Human Genetic Databases. Ashgate, Aldershot (in press)

Global Forum for Health Research [GFHR] (2002) The 10/90 Report on Health Research 2001-2002.

Harris J (2005) Scientific research is a moral duty. J Med Ethics 31: 242-248

HUGO Ethics Committee (2000) Statement on Benefit-Sharing. www.hugo-international.org/ PDFs/benefit.html. Accessed 10 September 2007

Juengst ET (1998) Groups as Gatekeepers to Genomic Research: Conceptually Confusing, Morally Hazardous, and Practically Useless. Kennedy Inst Ethics J 8: 183-200

Kaiser J (2002) Population Database Boom, From Iceland to the U.S. Science 298: 1158-1161

Knoppers BM, Fecteau C (2003) Human Genomic Databases: A Global Public Good? Eur J Health Law 10: 27-41

Macklin R (2003) Bioethics, vulnerability and protection. Bioethics 17: 473-486

Macklin R (2004) Double Standards in Medical Research in Developing Countries. Cambridge University Press, Cambridge

National Bioethics Advisory Commission [NBAC] (2001) Ethical and Policy Issues in Research Involving Human Participants. Vol. I. Report and Recommendations of the National Bioethics Advisory Commission. NBCA, Bethesda

Participants in the 2001 Conference on Ethical Aspects of Research in Developing Countries [Participants] (2002) Fair Benefits for Research in Developing Countries. Science 298: 2133-2134

Pullman D, Latus A (2003) Clinical trials, genetic add-ons, and the question of benefit-sharing. Lancet 362: 242-244

Simm K (2007) Benefit-Sharing and Biobanks. In: Häyry M, Chadwick R, Árnason V, Árnason G (eds) The Ethics and Governance of Human Genetic Databases: European Perspectives. Cambridge University Press, Cambridge

Schroeder D et al (2005) Sharing the benefits of genetic research. BMJ 331: 1351-1352

Schüklenk U, Kleinschmidt A (2006) North-South benefit sharing arrangements in bioprospecting and generic research: a critical ethical and legal analysis. Dev World Bioeth 6: 122-134

Schulz-Baldes A, Vayena E, Biller-Andorno N (2007) Sharing benefits in international health research. EMBO Rep 8: 8-13

United Nations Educational, Scientific and Cultural Organization (UNESCO) (2003) International Declaration on Human Genetic Data. http://portal.unesco.org/en/ev.php-URL_ID= 17720&URL_DO=DO_TOPIC&URL_SECTION=201.html. Accessed 10 September 2007

World Health Organization (WHO) (2002) Genomics and World Health. Report of the Advisory Committee on Health Research. WHO, Geneva

www.globalforumhealth.org/filesupld/1090_report_01-02/01_02_front_matt.pdf. Accessed 10 September 2007